Environmental Science, Engineering and Technology

www.novapublishers.com

Environmental Science, Engineering and Technology

Applications of AHP Methodology for Decision-Making in Cleaner Production Processes
Fisnik Osmani, PhD (Editor)
Atanas Kochov, PhD (Editor)
2022. ISBN: 978-1-68507-882-9 (Softcover)
2022. ISBN: 979-8-88697-012-8 (eBook)

Flow Diagrams Applied to Microalgae-Based Processes
Ihana Aguiar Severo, PhD (Editor)
2022. ISBN: 978-1-68507-742-6 (eBook)

Ecological Footprints: Management, Reduction and Environmental Impacts
Armano den Hartogh (Editor)
2022. ISBN: 978-1-68507-548-4 (Softcover)
2022. ISBN: 978-1-68507-606-1 (eBook)

Understanding Abiotic Stresses
Krishan K. Verma (Editor), Tatiana M. Minkina, PhD (Editor) and Vishnu D. Rajput, PhD (Editor)
2022. ISBN: 978-1-68507-508-8 (Hardcover)
2022. ISBN: 978-1-68507-552-1 (eBook)

An Innovative Approach of Advanced Oxidation Processes in Wastewater Treatment
Maulin P. Shah, PhD (Editor)
2021. ISBN: 978-1-68507-235-3 (Hardcover)
2021. ISBN: 978-1-68507-343-5 (eBook)

More information about this series can be found at
https://novapublishers.com/product-category/series/environmental-science-engineering-and-technology/

Binoy K Saikia
Editor

Atmospheric Aerosols

Properties, Sources and Detection

https://doi.org/10.52305/CXUG8701

NOTICE TO THE READER

Library of Congress Cataloging-in-Publication Data

ISBN: 979-8-88697-211-5

Published by Nova Science Publishers, Inc. † New York

Contents

Preface vii

Chapter 1 **Trace Elements and Rare Earth Elements in Aerosols** 1
Abel Inobeme, John Tsado Mathew, Charles Oluwaseun Adetunji, Alexander Ikechukwu Ajai, Stanley Okonkwo, Jonathan Inobeme, Mathew Adefusika Adekoya and Bamigboye Mutiat Oyedolapo

Chapter 2 **Microplastics in the Atmosphere: Emerging Air Contaminants** 23
Namita Yadav, Monika Vats, Dipanjana Chakraborty, Droupti Yadav and Kushagra Rajendra

Chapter 3 **Organic Compounds in Atmospheric Aerosols** 43
John Tsado Mathew, Charles Oluwaseun Adetunji, Abel Inobeme, Musah Monday, Elijah Yanda Shaba, Yakubu Azeh, Abulude O. Francis and Amos Mamman

Chapter 4 **Atmospheric Bioaerosol** 61
Shraddha Sanjiv Shirsat, Sharad Chandrakant Gangavane, Vijay Jagdish Upadhye, Babasaheb Shivmurti Surwase and Gauri Yogendra Kulkarni

Chapter 5 **An Introduction to Bioaerosols: Properties and Behaviour** 85
Nishi Srivastava

Chapter 6 **Source Apportionment of Atmospheric Aerosol Using Receptor Models: A Simple to Complex Approach** ... 117
S. K. Sharma

Chapter 7 **Source Identification of Aerosols Using Stable Carbon and Nitrogen Isotopic Composition** ... 133
S. K. Sharma and T. K. Mandal

About the Editor ... 145

Index ... 147

Preface

This book explores different aspects of atmospheric aerosols such as rare earth elements, trace elements, organic compounds, bioaerosols and microplastics emphasizing on their types, properties, sources, and analytical and source apportionment methods. All the chapters are authored by the experts in their relevant fields and contain up-to-date reference materials.

Chapter 1 - In recent times the quality of air in the atmosphere has significantly deteriorated as a result of the emission of Trace Elements (TEs), Rare Earth Elements (REEs) and other contaminants from various anthropogenic sources thereby constituting a serious problem of global concern. Trace elements and rare earth elements are ubiquitous and their concentrations in the atmosphere have risen over time due to human activities such as mining, combustions, smelting amongst others. Although some of these elements have a vital role as nutrients in trace amounts, some others have no known function in biological systems. The elements are also adsorbed on the particulate atmospheric matter transported in aerosol over long distances and are deposited on water bodies and other parts of the environment. As a result of the complex nature of particulate matter (PM) and aerosol particles, they have been identified as a vital indicator of the quality of air in a particular geographical area. An in-depth review of the elemental compositions of aerosol with regards to the presence and concentrations of trace and rare earth metals is paramount in understanding the sources and trends of pollution. This chapter reviews trace elements and rare earth elements in aerosols. It discusses the sources and distribution of these elements, factors affecting their concentrations and distribution, impact on humans and the environment, and recent reports on the presence and concentrations of these elements in the aerosol. Finally, an attempt is made on future trends of knowledge in this regard.

Chapter 2 - Plastics are miracle materials, but their extensive use is threatening the natural environment on which humanity and life critically depend, impacting aquatic and marine life, and the food chain and even detected at an unprecedented level in the atmosphere, from urban to the rural landscape and even in an indoor environment. Understanding of its atmospheric origin, life cycle, transport, deposition, impact, and interference with meteorology and climate is still at a nascent stage, possibly due to its extensive spread, diverse composition, and limited sampling and quantification methods. Microplastic forms of occurrence (fragments, shape, and nature) are aligned with the use pattern of plastics on the ground. Present study reviews contemporary knowledge on Suspended Atmospheric Microplastics (SAMP), including its physical & chemical characteristics, origin, release pathway, health exposure, movement (short term and long term), and deposition. This article also reviews and suitably highlights the sampling and detection methods in an indoor and outdoor environment. Microplastic occurrence is very diverse with an abundance and deposition across locations due to local and regional meteorology. To understand its impact, this article provides an overview of health impact and explores the correlation of its short- and long-term transport flux with weather & climate. Furthermore, the summary related to the findings on atmospheric microplastic properties like diversity, abundance, colors, shape, and size will provide a ready reference. The authors conclude with a holistic overview of an emerging area of microplastics in the atmosphere as an air pollutant and its linkages with human/ecosystem health and meteorology.

Chapter 3 - Organic compounds in the atmospheric aerosols and suspended particulate matter (PM) are a vital component of the atmosphere. Most of these compounds are volatile which affects their transport and distribution in the atmosphere. The presence of these compounds has become an issue of serious concern in recent times due to their toxic impact on human health and the environment at large, and their indispensable role as a driving force for various environmental issues. Polyaromatic hydrocarbons, polychlorinated compounds and some other components have been proven to be mutagenic and carcinogenic. Organic aerosols have also been reported to significantly affect climate and visibility. This chapter critically reviews organic compounds in the atmospheric aerosol. It discusses the occurrence and sources of these compounds, classes of organic compounds in aerosols, their impact on the environment and human health, recent reports on their distribution and concentrations in atmospheric aerosols. An attempt is also made at highlighting the future trends in this regard.

Chapter 4 - Air is a mixture of gases and other constituents, including biological organisms and cell debris. Recent pandemics and the rise of airborne ailments have highlighted bioaerosols vital role in combatting various infectious diseases, air pollution, and bioterrorism. The study of bioaerosols is also crucial for understanding the impacts of biogeochemical cycles, biodiversity, changes in ecosystems, air quality, and exposure in hospitals, forensic investigations, and industrial hygiene. Bioaerosols known as biological aerosols or organic dust are tiny particles in the range of 0.001 μm to 100 μm floating in the atmosphere with biological origins. They can be categorized as viable and non-viable. The viable bioaerosols include fungi, algae, viruses, and bacteria and non-viable bioaerosols include proteins, endotoxins, and dead cells. There has been tremendous development in the tools for collecting air samples and analyzing bioaerosols that can prove to benefit economic and ecological aspects. Despite this, there is a lack of standardization in regards to the maximum limit of concentration of atmospheric bioaerosols and the absence of uniformity in approaches for collection, analysis, and detection.This chapter deals to provide an overview of sources, types, transmission and transport mechanisms, collection, measurement, protection, and control methods, the significance of atmospheric bioaerosols, and future research needed to assess bioaerosols effectively.

Chapter 5 - The small solid or liquid particles suspended in the air are called aerosols, inhaled by a human while breathing. Even though being very small in size, aerosols play a vital role in human health and Earth's climate. Their sizes in the atmosphere vary from a few nanometers to micrometers, according to their origin and type. Aerosols present in the atmosphere have both natural and anthropogenic origins. Biological aerosols are usually called Bioaerosols. Bioaerosols are released into the atmosphere from the terrestrial or marine ecosystem. Bioaerosols are an essential component of the aerosol system in the atmosphere. They are less explored than their counterparts, such as dust aerosol, sulfate aerosol, black carbon aerosol, sea salt aerosol, etc. These bioaerosols consist of living and non-living organisms, including fungi, pollen, bacteria, viruses, and other biological fragments such as DNA fragments. Bioaerosols can transmit micro-organisms to humans, which can be allergic to human beings and cause severe illness. Bioaerosols enter the human body through inhalation and skin contact with air and various surfaces. Bioaerosols can cause an allergic response, toxic reactions, and infections in the human body. The bioaerosols became more relevant for study after a bio-terror attack in America in 2001 and influenza A H1N1 virus outbreak in 2009. The role of bioaerosols in transmitting the recent pandemic, Coronavirus disease 2019 (COVID-19), is also an important research area. World Health Organization (WHO) has announced the possibility of the spread of COVID-

19 through airborne transmission. Evidence indicates that aerosols contribute to the spread of COVID-19 under favorable conditions. The present chapter discusses the bioaerosol's origin, characteristics, behavior, distribution, and role in moderating human health and their possible role as agents in transmitting pandemics.

Chapter 6 - Thousands of articles have been published in the field of source apportionment of atmospheric aerosols (PM0.1, PM1, PM2.5, PM10, and TSP) in sub-urban and urban cities around the world, using various statistical tools and marker elements present in the aerosol to design mitigation strategies to improve global air quality. In this chapter, the authors have discussed the identifications and quantifications of possible sources of PM2.5 at a specific location of megacity Delhi, India using various statistical tools/receptor models [simple to complex: IMPROVE protocol, Enrichment Factor (EF), Principal Component Analysis (PCA), UNMIX, Positive Matrix Factorization (PMF)]. All these models identified the soil/crustal dust (SD), industrial emissions (IE), secondary aerosols (SA), vehicular emissions (VE), biomass burning (BB), fossil fuel combustion (FFC), and sea salts (SS) are the dominant sources of PM2.5 over the receptor site of megacity Delhi. This chapter discusses the effectiveness of various statistical methods used in source identifications and quantifications of atmospheric aerosol, which can help the stakeholders and authorities in identifying emission control initiatives to improve ambient air quality.

Chapter 7 - Recently radiocarbon and nitrogen analysis of aerosols have been used as a powerful technique to identify (decoding and tracking biogeochemical processes) the sources of aerosols around the globe. To characterize the sources of aerosols over the Indian region, stable isotopic ratios of carbon (δ13CTC) and nitrogen (δ15NTN) along with total carbon (TC) and total nitrogen (TN) have been reported in this chapter. δ13CTC and δ13TTN analysis of aerosols (PM2.5 and PM10) over the most polluted Indo-Gangetic Plain (IGP) region of India revealed that fossil-fuel combustion (FFC), biomass / wood burning (dung cake, C3 & C4 plant matter and wood, etc.), and secondary aerosols (SAs) are the important sources of aerosols over the IGP region. Levels of δ13CTC and δ13TTN detected in aerosols over the Himalayan region of India reveal biomass burning (crop residue and wood-burning), FFC and waste incineration as the important sources of aerosols. In this chapter, the authors discussed the composition of δ13CTC and δ15NTN in atmospheric aerosols (i.e., PM2.5 and PM10) and their possible sources.

Chapter 1

Trace Elements and Rare Earth Elements in Aerosols

Abel Inobeme[1, *], John Tsado Mathew[2], Charles Oluwaseun Adetunji[3], Alexander Ikechukwu Ajai[4], Stanley Okonkwo[5], Jonathan Inobeme[6], Mathew Adefusika Adekoya[7] and Bamigboye Mutiat Oyedolapo[8]

[1]Department of Chemistry, Edo State University Uzairue, Nigeria
[2]Department of Chemistry, Ibrahim Badamasi Babangida University Lapai, Niger State, Nigeria
[3]Applied Microbiology, Biotechnology and Nanotechnology Laboratory, Department of Microbiology, Edo State University Uzairue, Nigeria
[4]Department of Chemistry, Federal University of Technology Minna, Nigeria
[5]Department of Chemistry, Osaka Kyoiku University, Osaka, Japan
[6]Department of Geography, Ahmadu Bello University Zaria, Nigeria
[7]Department of Physics, Edo State University Uzairue, Nigeria
[8]Department of Chemical Sciences, Kings University, Odeomu, Nigeria

Abstract

In recent times the quality of air in the atmosphere has significantly deteriorated as a result of the emission of Trace Elements (TEs), Rare Earth Elements (REEs) and other contaminants from various anthropogenic sources thereby constituting a serious problem of global concern. Trace elements and rare earth elements are ubiquitous and their

[*] Corresponding Author's Email: inobeme.abel@edouniversity.edu.ng; abelmichael4@gmail.com

In: Atmospheric Aerosols
Editor: Binoy K Saikia
ISBN: 979-8-88697-211-5

concentrations in the atmosphere have risen over time due to human activities such as mining, combustions, smelting amongst others. Although some of these elements have a vital role as nutrients in trace amounts, some others have no known function in biological systems. The elements are also adsorbed on the particulate atmospheric matter transported in aerosol over long distances and are deposited on water bodies and other parts of the environment. As a result of the complex nature of particulate matter (PM) and aerosol particles, they have been identified as a vital indicator of the quality of air in a particular geographical area. An in-depth review of the elemental compositions of aerosol with regards to the presence and concentrations of trace and rare earth metals is paramount in understanding the sources and trends of pollution. This chapter reviews trace elements and rare earth elements in aerosols. It discusses the sources and distribution of these elements, factors affecting their concentrations and distribution, impact on humans and the environment, and recent reports on the presence and concentrations of these elements in the aerosol. Finally, an attempt is made on future trends of knowledge in this regard.

Keywords: aerosols, atmosphere, particulate matter, trace metals

Introduction

Environmental pollution by various organic and inorganic contaminants has become a pressing issue in recent times. Amongst the various contaminants released into the environment Rare Earth Elements (REEs) and Trace Elements (TEs) have become a subject of concern due to the environmental and health impacts of these elements. These substances are generated through various activities of man. These contaminants when released into the environment tend to partition within the particulate state and are distributed through atmospheric deposition onto water surfaces and the land. These substances tend to partition within the particulate state and are distributed through atmospheric deposition onto water surfaces and the land. Findings from different studies have revealed that atmospheric deposition is a major contributing factor to the contamination of the environment since this is responsible for the introduction of trace metals, rare earth elements (REEs), polycyclic aromatic hydrocarbons (PAHs), polychlorobiphenys (PCBs), and various compounds of nitrogen into the atmosphere (Balaram et al., 2019; Dai et al., 2016 and Mohan et al., 2021).

More recently, trace metals and REEs have constituted one of the major classes of contaminants in aerosols.This has seriously affected the atmosphere thereby posing a serious issue of environmental concern. Due to the global rise in the usage of trace elements and rare earth elements, there has been an increase in the emission of these elements into the atmosphere.Several studies have documented that the primary source of urban environmental contamination with high trace and rare metals is atmospheric deposition (Fabio et al., 2016; Yan et al., 2020 and Gapmlkar et al., 2019). Some of the major reports have also identified dry deposition as the primary contributing factor to the movement of airborne toxic materials. Particles dispersed through dry deposition are made of comparably large aerosols, bigger than 10 microns in their dimension (Stewart, 2020).On emission of these elements, they pass through long-range transport processes through various pathways and are finally deposited into land surfaces as well as water bodies. This has contributed consequently to the contamination of agricultural lands as well as the aquatic ecological systems. Although in some cases, some of these elements act as nutrient sources to the ecosystem, they could, however, constitute significant harm when their concentrations exceed specific limits, and in some cases, most of the elements do not have any known physiological roles in biological systems. The accumulation of these metals in the tissues of organisms as they are conveyed through the food chain has serious deleterious effects on organisms as this can result in damage of vital body organs in humans and other animals (Ali et al., 2019). Due to the negative health impact of these elements, the study of the elements as well as the role they play in the contamination of the atmosphere, therefore, become paramount. Aside from their impacts on human health at the regional scale, some of the mares are capable of acting as co-limiting or limiting nutrients for the productivity of the surface ocean ecosystem.Also, some of the elements are capable of acting as toxicants thereby bringing about a reduction in primary productivity. The atmospheric flow systems which have a high amount of REEs could significantly modulate and influence surface ocean biogeochemical processes. It is therefore pertinent to quantify their contents in various ecosystems and evaluate their atmospheric sources as well as distributions on the temporal and spatial scale (Balali-Mood et al., 2021).

The chemical family of elements known as the REEs also described as the lanthanides are made of 14 metallic elements that are naturally occurring and range from lanthanum (La) to lutetium (Lu), however, excludes the element Promethium (Pm) that does not occur naturally in a stable form. Some literature has also included Scandium (Sc) and Yttrium (Y) in this category of elements. These elements have unique chemical properties due to their peculiar oxidation states, electronegativity properties and electronic structures. Rare Earth Elements are classified into three major groups: the light group of rare earth elements which range from Lanthanum (La) to Neodymium (Nd), the medium group, ranging from Samarium (Sm) to Dysprosium (Dy) and the heavy rare earth elements from Holmium (Ho) to Lutetium (Lu). Despite their unique similarity in physicochemical properties, the light REEs show higher solubility and a lesser tendency of complex formation when compared to heavy REEs (McLennan, 2018). The REEs are different from other elements in that when observed from the anatomical point of view they look inseparable from one another in that they show very close chemical properties. Each of them is however unique concerning its magnetic properties and electronic properties. They also show unique applications depending on the chemical nature of the element. The exploitation and utilization of rare earth minerals and trace elements on large scale have no doubt contributed to the significant rise in the concentration of the REEs in water, soil and air. Within the last few years, studies have been carried out focusing on the concentration as well as distribution of REE in water and soil (Mehmood, 2018). There are however fewer studies and reviews on the concentration of these metals in the atmosphere which is a result of the limitations of analytical processes. The determination of the elemental compositions of ambient particulate matter is vital in the evaluation of the possible sources as well as effects on the environment. Several studies in this regard have focused on some organic compounds and other volatile compounds with sparse information on rare earth metals (Manisalidis et al., 2020).This chapter reviews trace elements and rare earth elements in aerosols. It highlights the sources and distribution of these elements, factors affecting their concentrations and distribution, impact on humans and the environment, and recent reports on the presence and concentrations of these metals in the aerosol. Finally, an attempt is made on future trends of knowledge in this regard.

Sources and Distribution of REEs and TEs

Except for scandium, other REEs have densities greater than that of iron hence they are produced through supernova nucleosynthesis or by the s-process within asymptotic giant stars.Naturally, the spontaneous disintegration of uranium with an atomic mass of 238 gives rise to a small amount of promethium that is radioactive whereas most promethium is produced synthetically during nuclear reactions. Significant amounts of the oxides of rare earth metals are generated during tailings. As a result of the increasing costs of the rare earth elements, their extraction has become viable economically (Vlasov et al., 2020).

More recent investigations on an atmospheric deposition by different authors have made immense input to the knowledge of the various sources of TEs and REEs as well as their spatial and temporal deposition (Balalam, 2018 and Wiklund et al., 2020). One of the major contributions in this regard is that high concentrations of the coarse aerosols are commonly found far away from their sources when compared to what would have been obtained if there was no re-suspension. This brings about regional than local effects of the specific sources. This has given rise to a potential hypothesis that when a possible mechanism of resuspension is envisaged, typically wind or traffic, there is the incessant lower layer of pollutant laden aerosols (Wiklund et al., 2020). REEs when released into the atmosphere from a source are deposited firstly within the locality of the source, then are later re-suspended, deposited again and then finally re-suspend.The particulate substances that are released during the re-suspension known as the fugitive dust have been identified as the primary source of atmospheric trace metals. These atmospheric processes continue until these substances are finally washed by rainfall or can no longer re-suspend (Crocket et al., 2018).

The rare earth metals are not found in nature as free native metals like copper, gold and silver which is a result of their reactivity. They are mostly found as ores with other related mineral components. They are present in different groups of minerals such as carbonates, phosphates, oxides and silicates. They do not however fit into most mineral morphology hence found only in a few geological environments. The major economic source of the REEs minerals includes loparite, bastnaesite, lateritic clays and monazite. Naturally, there are more than 250 minerals that have REEs as major components in them. More recently, however, the REEs contents in the atmospheric dust have been employed for the assessment and tracing of the possible origin of the environmental harmful materials. The REEs occur naturally in minerals and rocks as accessories and are also found in certain types of deposits such as igneous deposits. It is estimated that the mean concentration of REEs in the earth ranges from 130 ug/g – 240 ug/g in the crust and this is significantly greater than those of the other elements (Balaram, 2018).

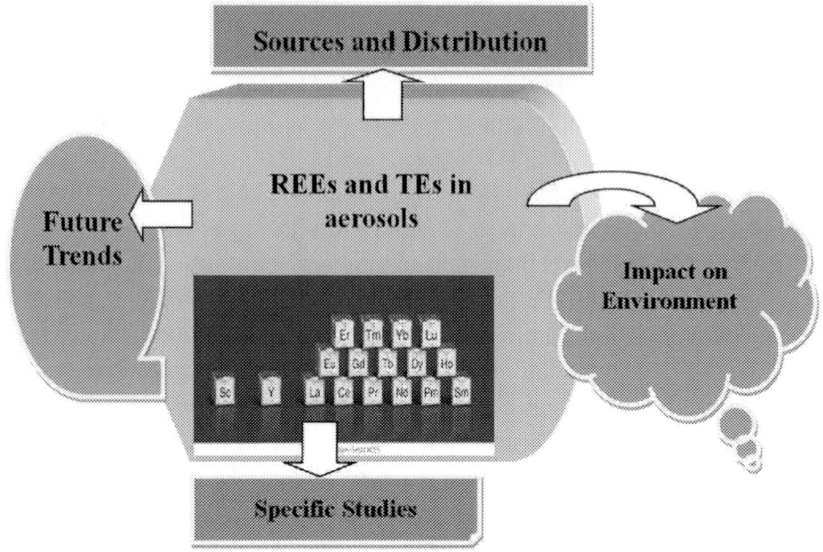

Figure 1. REEs and TEs in aerosols, sources, distribution, impact on the environment and specific studies.

REEs are mined from their natural deposits and then used in various industries in the production of various materials such as the light-emitting diode, screens of television, bulbs, nuclear batteries, optic glass, tubes for X-ray and X-ray machines.The significant increase in the concentrations of REEs in the environment has been attributed to the indiscriminate discharge of incineration of electronic waste. The extraction of the REEs from electronic waste has been enhanced due to the advancement in the area of recycling technology. For instance, in Japan, it has been reported that a total of 300,000 tons of these elements are found in electronic wastes. Similarly, in France, the production of 200 tons of REEs has been carried out by two factories from used batteries, fluorescent lamps and magnets. Coal and related by-products are also possible sources of REEs. Mining of REEs, production of phosphorus-based fertilizers and the application of fertilizers enriched with REEs can bring about environmental pollution due to these elements (Binnemans et al., 2018). These elements are also capable of speciating depending on the condition of the environment such as soil. Also depending on their bio-availability to plants, they could be absorbed within the plants, accumulated and later transported across the food chain.

Various minerals such as apatite, zircon, monrite which are sedimentary rocks have been known to have the potential of concentrating and fractionating

REEs. Several studies have also documented that these minerals are not capable of controlling the distribution of the rare earth elements within the sediment (Cavalcante et al., 2014). Studies have also documented that the abundance of REEs is also affected clay minerals, whereas some suggested that the major carrier REE is the kaolinite. In addition, authigenic mineral states concentrated in ultra-fine and fine compositions are the most reliable in addressing the fractionation of REEs induced by hydrothermal or diagenetic phenomena. The studies carried out by some of the authors show that the accessory minerals elements do not affect significantly the REE composition in the shales and that the patterns of distribution of the size fractions analysed are made up majorly if not exclusively of combined components of illite-smectite (Abbott et al., 2019; Turetta et al., 2021).

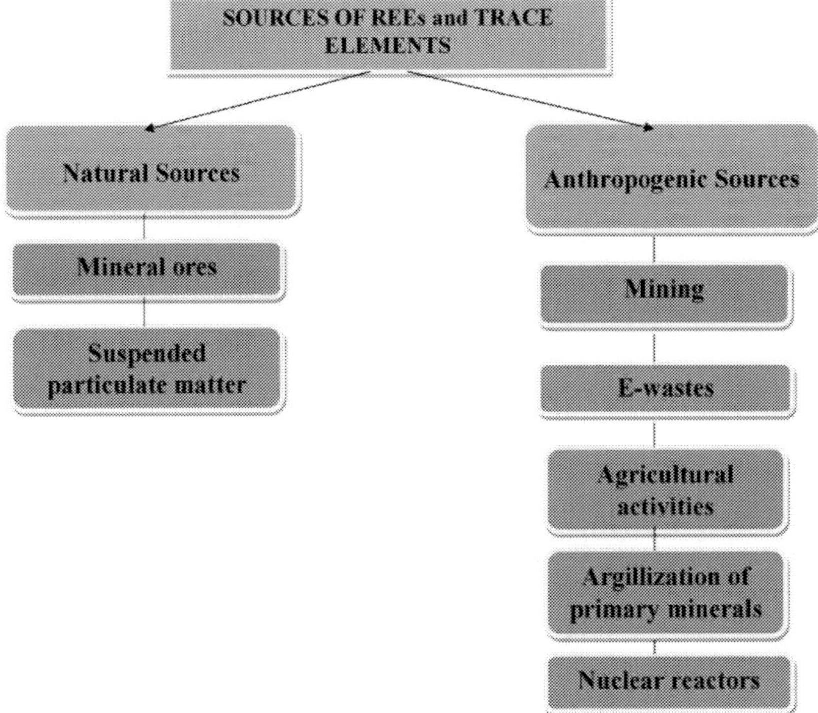

Figure 2. Sources of REEs and TEs.

Table 1. Physicochemical properties and uses of REEs

Element	Atomic number	Abundance (part per million)	Uses
Scandium (Sc)	21	22	Component of alloys for aerospace and as radioactive tracers in oil industries. Also as a component in the production of halide lamps
Yttrium (Y)	39	33	In lasers,refractory components, tooth crown, electro-ceramicsand fuel cells
Lanthanum (La)	57	39	Glass production,catalyst in refineries andelectrodes in the battery
Cerium (Ce)	58	66.5	Coating agents for turbine blades, catalyst in refineries, colouring agents in ceramics and glass
Praseodymium (Pr)	59	9.2	Colouring agents in enamel and glass production. Component of the optical amplifier
Neodymium (Nd)	60	41.5	Electric automobile components,capacitors, lasers and magnets production
Promethium (Pm)	61	1×10^{-15}	Batteries in nuclear plants and production of paint.
Samarium (Sm)	62	7.05	Reactors in nuclear plants, lasers, magnets, and control rods.
Europium (Eu)	63	2	Relaxation agents in NMR, lasers, lamps (fluorescent and mercury vapour)
Gadolinium (Gd)	64	6.2	Additive in alloying, detectors in scintillators, lasers and glasses.
Terbium (Tb)	65	1.2	Component in sonar systems,fuel cell stabilizers and fluorescent lamps
Dysprosium (Dy)	66	5.2	Additives in magnets, components of alloy and hard discs.
Holmium (Ho)	67	1.3	Component of optical spectrophotometers and lasers
Erbium (Er)	68	3.5	Additives in vanadium steel, lasers and optical fibres
Thulium (Tm)	69	0.52	Lasers, halide lamps and X-ray devices
Ytterbium (Yb)	70	3.2	Additives in stainless steels, lasers, and stress gauge
Lutetium (Lu)	71	0.8	Detectors, catalysts in oil refineries and bulbs for LED light.

Several elements in $PM_{2.5}$ including the toxic ones like Arsenic, chromium, lead and cadmium has been documented (Binnemans et al., 2018). Also, the platinum class elements like Pt, Rh and Pd have been focused on $PM_{2.5}$ with regards to anthropogenic discharge from automobiles. However, studies on REEs in $PM_{2.5}$ are not much in comparison. REEs show similarities in their physicochemical properties and are seen to show high resistance to the process of fractionation within a supra crustal environment. The distribution pattern and composition of the REEs are essentially constant; hence they have been widely employed in the aspect of tracing the provenance of the sediments as well as the atmospheric particulate matter. The physical and chemical parameters related to the REEs are related strongly to their compositions. Contamination of air due to the REEs in particulate forms has also been linked to the burning of fossil fuel around the vicinity of refineries. This is due to the high enrichment of the catalyst used for the oil cracking process with REEs (Ali et al., 2019).

FactorsAffecting Distribution of REEs and TEs

Several contributing factors affect the dispersal of the particulate pollutants into distance environment, which include humidity, wind direction, wind speed, dispersive conditions within the atmosphere, and scavenging tendencies of rain, geographical position of the areas, nature of soil covering, closeness to the coast, topography and season. The morphology of the particulate matter could also affect their re-suspension, transport mechanism and extent of toxicity. Most of the particulate matters dispersed through the air are produced from the weathered products of the various geological processes which are then conveyed by the air in motion. The transport mechanisms of trace and REEs as well as their route of distribution into different components of the environment is commonly affected by photochemical and surface phenomena and then combined with masses of air which have various kinds of aerosols including substances from the combustion of biomass and other materials generated through human activities (Briffa et al., 2020).

Impact of Trace and REEs on the Environment and Humans

REEs and trace elements in aerosols have been connected with several harmful health impacts, negative effects on regional and local climatic conditions, transport of elemental nutrients within the biosphere and distribution of anthropogenic pollutants (Kolawole et al., 2021). The trace metal contents of atmospheric particulate matter have been vital in the provision of details with regards to the origin of dust particles as well as their deleterious effects. Even though the contents of the REEs within the $PM_{2.5}$ are low, they are capable of bringing about serious health to humans and other animals. Hence the deleterious environmental and health impacts of some trace elements and REEs in the atmosphere, as well as other particulate matter suspended, have attracted much concern in recent times.Atmospheric distribution and deposition of the REEs and trace elements also affect the ecosystem. In other to ensure environmental protection it is vital to assess the input of these elements from various sources on different components of the environment (Turetta et al., 2021).

Impact on Vegetation

The presence of trace and REEs has resulted in the contamination of various components of the environment most especially the soil, thereby interfering with the nature of the vegetations in the polluted areas. Contamination can bring about the inhibition of the growth of plants as well as a decrease in the production of chlorophyll which is vital for the process of photosynthesis. Some plants are capable of absorbing and retaining these elements. The potential of the plants to absorb these elements however depends on the nature of the elements (Chibuike andObiora, 2014).

Effect on Aquatic Organisms

The large scale dispersion of other several trace elements in atmospheric are making provision for insight into the extent of atmospheric sources, the role played by trace elements in the phenomenon of biogeochemistry and the tendency for limitations in the growth of phytoplankton by various metals such as manganese, zinc and cobalt and other trace elements (Merchant and Helmann, 2014). It has also been documented that the bioavailability o cobalt zinc is capable of limiting the potential of plankton to make use of organic phosphorus through the alkaline phosphates which need these elements as co-factor. The available data based on observations for manganese has aided the formulation of the initial model of the global cycle of marine manganese. The effect of REEs depositions from the atmosphere is dependent on the current nutrient status of the surface water receiving the deposit.It has been established that deposition of atmospheric dust brings about an increase in the content of some trace metal ions within the surface water of the tropical Atlantic (Balaram, 2019).

Animal Health

Studies that involve the exposure of experimental rats to different compounds of cerium documented the accumulation of these elements majorly in the liver and lungs. In some cases, the REES and some TEs have been introduced into the feeds of livestock to increase the production of milk and body mass (Cao et al., 2020). Some REEs have been introduced for the improvement of the bodyweight of pigs. Different investigations have however reported the need for considering dose-response when considering the positive impact versus toxicity of these elements (Cassee et al., 2011 and Modrzynska et al., 2018). In some other available reports, animals found around the vicinities of areas contaminated with these elements have been observed with failure in organs and entire systems (Wang et al., 2013 and Rai et al., 2019).

Effect on Humans

Particulate matter contamination has been identified as one of the most tedious challenges for resolution in the area of human diseases especially those that are airborne. REEs are complex classes having different physicochemical properties and concentrations within the environment. As a result of this, as well as the few studies in this area, it has been tough determining the safe exposure levels in humans. Some available research has focused on the aspect of risk assessment based on divergence in the background contents as well as exposure. Some of the documented health impacts in humans include dental loss, respiratory diseases and even death (Rai et al., 2019).

REEs have the potential of accumulating in human tissues and eventually deposited in various organs through ingestion of food substances, breathing in of ambient particulate matters as well as contact with the skin. Numerous investigations have reported that some REEs have vital functions at trace concentrations and harmful impacts at high contents. It has also been documented that high concentrations of REEs could bring about serious detrimental effects on vital organs. Some of the adverse effects associated with high concentrations of the REEs include inhibition of growth, toxicity effect on vital organs as well as cytogenetic impacts. Research-based on occupational effects has also shown a long term health risk of pneumoconiosis among occupational workers due to exposure to the high concentration of metals (Xing et al., 2016). These elements are capable of accumulating in vital organs such as the lungs, lymphatic nodes amongst others. Pollution of the aerosol due to heavy and rare earth metals is affected by several aerodynamic parameters. A detailed comprehension of the adverse impacts of the REEs on human health requires a thorough investigation of the fractionation of the atmospheric REEs in size-resolved aerosols. Pollution associated with particulate matters in the atmosphere has a serious effect on visibility as well as air quality (Malhotra et al., 2020).

Standards on Trace Elementsand REEs in Aerosols

Rare earth elements are naturally found in trace concentrations in the environment. Mines are mostly found in countries with very low social and environmental standards thereby engendering environmental pollution. Within the proximity of industrial and mining sites, the concentration of the REEs increases to several times the background contents (Aide and Aide, 2012). The evidence that high concentrations of rare earth and heavy metals have deleterious health impacts led to the WHO revision of the quality for air guidelines and the imposition of a global permissible standard for the PM_{10} content in the inhaled air. However, no serious consensus has been reached concerning the fraction of the size that will determine the pattern of deposition within the respiratory tract, and also the chemical constituents that are highly implicated with regards to bioreactions. The key components that have been suspected to be of potencies include the trace elements which have a high bio reactivity hence play a role in human degenerative diseases.Even though the contents of these metals are usually very high in urban air, just a few among them have been legislated. For example, the WHO set standards and guidelines for various elements (Cd: 5 ngm^{-3}; Pb: 500 ngm^{-3}; Mn: 150 ngm^{-3} and V: 1000 ngm^{-3}) for daily values. Similarly, the European Union has also set standards for some metals (Pb: 500 ng^{-3}, Cd: 5 ng^{-3}, Ni: 20 ng^{-3} and As: 6 ng^{-3}). Metalliferous atmospheric particulate matters usually prefer to concentrate in the fractions made of particulate matter (PM_1 and $PM_{2.5}$) and are found in sizes lower than 1 μm which does not only bring about a large surface area that makes them available for reactions within human fluids but also enhance their potential to be dispersed several kilometres away from their origin. This accounts for the movement of trace and rare elements contaminants generated in the urban environment to rural and remotest areas thereby polluting the air in these areas (Asl et al., 2017).

Studies on TEs and REEs in Aerosols

Mahowald et al. (2018) in their work reported that the dissolution of metal from particulate aerosol deposition from the atmosphere into the ocean is paramount in enhancement and inhibition of the growth rate of phytoplankton and the modification of the community structure of the plankton thereby affecting the biogeochemistry of the marine ecosystem. They did a comprehensive review covering the. They took into consideration the deposition of the aerosol on the short time intervals during which the phytoplankton response to the metallic causes and impacts of the leaching processes of several trace metals from anthropogenic and natural aerosol inputs together with the longer time scale during which the aerosol affects the biogeochemical processes.

In a related study, Turetta et al. (2021) investigated the annual variability of the fractions of trace elements that are water-soluble as well as the rare earth elements. They collected airborne particulate matter samples for a period of one year (2018 – 2019). A total of six aerosol fractions with their specific dimension were collected through the use of a multistage Andersen device for a thorough comprehension of the regional and local circulation pattern, to identify the source of the various inorganic tracers from specific anthropogenic and natural sources. They also utilized a factor analysis on the set of data to evaluate the potentials of certain inorganic tracers serving as indicators for specific regimes of circulation.

Also, Jickells et al. (2016) did a review work on the atmospheric contribution of trace metals and mineral nutrients to the ocean with regards to the GEOTRACES programs and thereby made a provision for novel data generated from Atlantic GEOTRACES cruises. They took into consideration the deposition of mineral dust as well as trace metals, deposition patterns of other trace elements from natural and anthropogenic sources and nitrogen deposition in the ocean. They also evaluated the extent of the solubility of the various elements.

Khondokeri et al. (2013) investigated rare earth elements and trace elements in volcanic ash, sediments, road dust and aerosol filters from selected cities in Johannesburg. They observed that all the sites could act as potential sources of atmospheric aerosols. They utilize the data on trace and rare earth element concentrations as well as the isotopic details for the characterization of the various sources of natural and anthropogenic aerosols. They concluded that the provenance information would be vital in the interpretation of the first detailed geochemical set of data on trace elements, rare earth elements and isotopic compositions for the aerosol filters obtained.

In another study, Radomskaya et al. (2018) investigated the elemental composition, the pattern of distribution and fractionation of the rare earth elements in the suspended and dissolved forms of precipitation in the atmosphere. In their work, they analyzed samples of snow within the urban area. The findings from the electronic microscopy revealed a high amount of the mineral elements in the aerosols and identified the major source of the pollution was as a result of the emission from power and heat plants as well as boiler houses. Some geochemical anomalies that differed were observed in the snow covering of the city. The concentrations of the REE in the solid phase were relatively higher when compared to those of the liquids and most of the samples in the snow liquid-solid phase revealed poor fractionation.

Yan et al. (2020) observed that the chemical profiling of $PM_{2.5}$ fugitive and aerosol dust in different cities have been well studied by various researchers. They however identified that the provenance and specific implications of the REEs as well as some isotopes such as Nd and Sr have not been well documented. They collected $PM_{2.5}$ fugitive and aerosols from Nanchang, in China and analyzed the specific and provenance impacts using REEs and the isotopes of Sr-Nd. They reported higher REEs in the $PM_{2.5}$ aerosols in comparison to the fugitive samples of dust from the road. There was an indication of REEs enrichment and some negative anomalies in both the fugitive sample of dust and $PM_{2.5}$ aerosols; this was deduced from the chondrite normalized distribution of REEs as well as the REEs distribution trend. They observed that the REES present in the $PM_{2.5}$ dust samples was significantly impacted by various activities which include steelworks, combustion of coal and construction using cement as well as the background soil contribution. They concluded that the REEs emerged from both local and non-local sources.

In a different study, Turetta et al. (2016) focused on organic compounds that are water-soluble and elemental constituents in the particulate airborne matter to assess the distribution and fractionation of the REES, trace elements and the water-soluble constituents of the organic compound. Collection of samples was done in the Svalbard Islands and the compositions of the REEs and trace elements were determined with the aim of identifying the reliable tracers of characteristic sources, which could be vital in the economical control of air pollution. The content of the trace elements and REEs were revealed to be due to long-distance dispersion from distant areas. Findings from the study provide detailed information on the circulation of aerosols at the global scale as well as the impact of anthropogenic factors on the compositions of the aerosols.

Kolawole et al. (2021) investigated the sources of atmospheric contamination using the pattern and distribution of REEs. They collected a total of twenty-five samples of atmospheric dust in the area investigated. The area sampled includes dumpsite, industrial area, traffic area, remote area and residential area all within Ibadan, Nigeria. For effective comparison, they also took samples from the topsoil and granite rock samples. The determination of the concentration of the REEs was done using the ICP-MS. The value of La/Ce is the ratio of lanthanum to cerium in some of the locations within the industrial area was found to be 1.5; in the traffic, the area was also 1.5 while the dumpsite was 1.1. These were found to be higher than that of the soil (0.2) as well as the upper continental crust (0.5).

Similarly, Gapmlkar et al. (2019) assessed major elements such as iron, potassium, aluminium, magnesium, calcium as well as some trace elements such as lead, zinc, nickel, chromium, copper, and manganese present in atmospheric around three different sites in Goa, Eastern Arabian Sea. The assessment was done within the winter and summer of 2015. They reported a significant temporal and spatial disparity in the content of P_{10} mass, trace elements and crustal elements present in the samples. Mineral dust composition was determined through the application of the diagnostic crustal elementary ratio. Enrichment factor assessment was also employed for the determination of the concentrations of the different trace elements investigated. Based on the enrichment factor, they were categorised into two groups. The first group was made of copper, chromium and lead having $10<EF< 100$ in comparison to values of the continental crustal, which therefore suggested a significant input from natural origin. The second category consist of nickel and zinc which showed a significantly high enrichment factor that was higher than 100. This shows the contribution from anthropogenic sources. Source identification of the trace elements was further done using the principal component analysis, which revealed that the primary input of the trace elements was from anthropogenic origin.

Hu et al. (2019) worked on the geochemical properties and origin of the REEs in $PM_{2.5}$ in Quanzhou City. Samples of $PM_{2.5}$ were obtained from three different areas during different seasons. Analysis was then carried out on the distribution and compositions of the REEs. The major provenance of the different elements was also assessed using the use of ternary diagrams. Also tracing was done to verify the sources of the $PM_{2.5}$ through the utilization of the diagram of the Nd isotopic ratios and the REEs characteristics. The contents of the total REEs, total HREEs, and the light REEs were observed to show variations within the different seasons investigated. A significant correlation was observed between the total REEs in the $PM_{2.5}$ and the mass concentration during the spring, winter and summer.

Ferrat et al. (2011) in a related study utilized an advanced geochemical framework provenance tracer for the assessment of the circulation pattern of the atmosphere as well as the extent of variability in the central Asia region. They combined information gathered from major and trace elements as well as the mineralogical details of the Chinese, Tibetan and Indian dust sources. They observed that clay minerals, feldspars and quartz are the primary components of all the sources investigated, with calcite contents showing high variability as revealed in the concentration of the calcium oxide.

Also, Agnan et al. (2014) evaluated the atmospheric deposition of the REEs through the use of mosses, lichens together with herbarium samples collected from some forests in France.A comparison of the REEs pattern of distribution in the organisms, as well as the bedrocks, revealed a localised homogeneity influence within the dust particles emerging from the bedrock and the soil weathering that was entrapped by the mosses and lichens.

Fabio et al. (2016) in their work reported a first attempt in determining and interpreting the behaviour of metals and REEs around the polar aerosol at a high temporal level. They collected samples of PM_{10} daily using Teflon filters. Inductively coupled plasma MS was then employed for the chemical analysis and quantification of the trace and rare earth metal concentrations. They utilize the results for the detection of the source, transport phenomenon and depositional impacts of the anthropogenic and natural particulate getting to the Arctic from the industrialised area. They concluded that the REEs could be used as a vital fingerprint in the soil for assessing sources of REEs.

Dai et al. (2016) in their study investigated the potential source of the REEs together with the atmospheric signature in samples of ambient particulate matter that were collected background (ZWY) and urban (ZH) sites within a period of one year (2011 to 2012). The mean sum concentrations of the REEs for $PM_{2.5}$, PM_{10} and TSP were found to be 2.24, 9.37 and 11.98 ng/m3 in ZH, and these were found to be higher than those detected at ZWY. The distribution pattern of the REEs concerning sizes revealed that they are fractionated significantly into about 50% coarse particles within $PM_{2.5}$ to PM_{10}. It was also observed that Sm, Ce and La were of higher enrichment in the PM with regards to each other and the other rare earth metals. Contamination by Ce was also observed to be of higher prominence within the area investigated and this was found in a relatively higher amount within the particulate matter of finer sizes. The Enrichment factor obtained from this study shows that Nd, Sm, Ce ad La originated from the combination of natural and anthropogenic emissions.

Spickoya et al. (2011) did a related study in which a comparison of the REEs atmospheric signature in different compartments of the ecosystem was done around the vicinity of the LesniPotok catchment (LP) in the Czech Republic. Investigation of the REEs signature was done within a period of two years (2004 to 2006) in lichens, rainwater and tree foliage for some selected elements which include potassium, calcium, iron, aluminium, silicon, sulphur phosphorus, magnesium and manganese. The REEs were observed to increase in a particular sequence: lichens >beech leaves>spruce through fall> beech through fall > bulk precipitation. The crustal source of the REEs in a large amount of the precipitation was discovered by the high correlation with the lithogenic elements, through seasonal disparity in the concentration of REEs.

A detailed summary of the findings from various studies on TEs and REEs is presented in Table 2 below.

Conclusion

This chapter has discussed REEs and TEs in aerosols. Various sources, distribution, environmental impacts and factors affecting their dispersion were highlighted. Recent research has suggested paramount nutrient and toxicity impacts from many aerosols. There is a need for continuous advancement in the methods for the determination of trace elements and REEs.More robust techniques of observing and monitoring depositions, both dry and wet, would enhance a thorough understanding of elements in the aerosols. At the moment, inferences on deposition rate are drawn from ocean and atmospheric concentrations or the application of ocean sediment deposits which times are located about 1000m beneath the surface. A similar approach involves the utilization of metal isotopes for the provision of specific constraints on the origin and sinks of elements in the ocean and atmosphere. It is also paramount to put in place a time series of long term input focusing on stations for the measurement of aerosols as well as evaluation of their impacts on the water bodies. Studies that focus on modelling and big data could be incorporated for the provision of the database on atmospheric and ocean sources as well as the three-dimensional approach to modelling. There is a need for detailed investigation into various elemental compositions of the aerosols since most of the available studies have been focused on iron. Such studies should also be extended beyond nutritional perspectives but rather the toxicity due to these elements.

Table 2. Summary of studies on REEs and TEs in aerosols

Scope of investigation	Findings	References
Causes and impacts of the leaching processes of TEs from anthropogenic and natural aerosols	The dissolution of TEs and REEs affect the biogeochemistry of the marine ecosystem	Mahowald et al. (2018)
Advance geochemical framework provenance tracer for the assessment of the circulation pattern of the atmosphere.	Clay minerals, feldspars and quartz are the primary components of all the sources investigated.	Ferrat et al. (2011)
Annual variability of fractions of TEs and REEs in aerosols	Inorganic tracers such as REEs could act as indicators for specific regimes of circulation.	Turetta et al. (2021)
Determination and interpretation of the behaviour of metals and REEs around the polar aerosol	REEs could be used as a vital fingerprint in the soil for assessing sources of REEs	Fabio et al. (2016)
Evaluation of the atmospheric deposition of the REEs through the use of mosses lichens.	Localized homogeneity influence the composition of the dust particles emerging from the bedrock	Agnan et al. (2014)
Atmospheric contribution of TEs and mineral nutrients to the ocean.	Mineral dust is a vital source of TEs and REEs. The TEs also vary in their extent of solubility.	Jickells et al. (2016)
REEs and TES in volcanic ash, sediments and aerosols	Provenance information would be vital in the interpretation of the detailed geochemical set of data on trace elements.	Khondokeri et al. (2013)
REEs composition and pattern of distribution in atmospheric precipitation.	The concentrations of the REE in the solid phase were relatively higher when compared to those of the liquids.	Radomskaya et al. (2018)
Analysis on the specific and provenance impacts using REEs and the isotopes of Sr-Nd.	The REEs emerged from both local and non-local sources.	Yan et al. (2020)
Distribution and fractionation of the REES, TEs and the organic compounds	The content of the TEs and REEs were revealed to be due to long-distance dispersion from distant areas.	Turetta et al. (2016)
Sources of atmospheric contamination using the pattern and distribution of REEs	The ratio of lanthanum to cerium in some of the locations within the industrial area was found to be higher than that of the soil (0.2) as well as the upper continental crust (0.5).	Kolawole et al. (2021)
TEs in atmospheric around three different sites in Goa, Eastern Arabian Sea.	Source identification of the TEs revealed that the primary input of the TEs was from anthropogenic origin.	Gapmlkar et al. (2019)
Geochemical properties and origin of the REEs in $PM_{2.5}$ in Quanzhou City	A significant correlation was observed between the total REEs in the $PM_{2.5}$ and the mass concentration.	Hu et al. (2019)

References

Abbot, N., Lohr, S., and Trethewy, M. (2019). Are Clay Minerals the Primary Control on the Oceanic Rare Earth Element Budget? *Front. Mar. Sci.*, 20 https://doi.org/10.3389/fmars.2019.00504.

Ali, H., Khan, E., and Ilahi, I. (2019). Environmental Chemistry and Ecotoxicology of Hazardous Heavy Metals: Environmental Persistence, Toxicity, and Bioaccumulation. https://doi.org/10.1155/2019/6730305.

Asl, F., Leili, M., Vaziri, Y. and Ferrante, M. (2017). Health impacts quantification of ambient air pollutants using AirQ model approach in Hamadan, Iran Air pollution assessment. *Environmental Research*, 161, 114-121. DOI: 10.1016/ j.envres.2017.10.050.

Balali-Mood, M., Naseri, K., Tahergorabi, Z., Khazdair, M., and Sadeghi, M. (2021). Toxic Mechanisms of Five Heavy Metals: Mercury, Lead, Chromium, Cadmium, and Arsenic. *Front. Pharmacol.*, https://doi.org/10.3389/fphar.2021.643972.

Balaram, V. (2018). *Rare earth elements: A review of applications, occurrence, exploration, analysis, recycling, and environmental impact.* doi.org/ 10.1016/j.gsf.2018.12.005.

Balaram, V. (2019).Rare earth elements: A review of applications, occurrence, exploration, analysis, recycling, and environmental impact. *Geoscience Frontiers*.10, 4, 1285-1303https://doi.org/10.1016/j.gsf.2018.12.005.

Benckiser, G. (2012). *Rare Earth Elements: Their Importance in Understanding Soil Genesis*. doi.org/10.5402/2012/783876.

Binnemans, K., Jones, P.T., Müller, T., et al. (2018). Rare Earths and the Balance Problem: How to Deal with Changing Markets?. *J. Sustain. Metall.*, 4, 126–146. https://doi.org/10.1007/s40831-018-0162-8.

Briffa, J., Sinagra, E., and Blundell, R. (2020). Heavy metal pollution in the environment and their toxicological effects on humans. *Heliyon*, 6, 9, e04691 https://doi.org/10.1016/j.heliyon.2020.e04691.

Cao, B., Wu, J., Xu, C., Chen, Y., Xie, Q., Ouyang, L., and Wang, J. (2020). The Accumulation and Metabolism Characteristics of Rare Earth Elements in Sprague–Dawley Rats. *International Journal of Environmental Research and Public Health*, 17, no. 4, 1399. https://doi.org/10.3390/ijerph17041399.

Cassee, F. R., van Balen, E. C., Singh, C., Green, D., Muijser H., and Weinstein (2011). Exposure, health and ecological effects review of engineered nanoscale cerium and cerium oxide associated with its use as a fuel additive. *Crit Rev Toxicol.*, 41, 213–229. 10.3109/10408444.2010.529105.

Cavalcante, F., Belviso, C., Piccarreta, G., and Fiore, S. (2014). Grain-Size Control on the Rare Earth Elements Distribution in the Late Diagenesis of Cretaceous Shales from the Southern Apennines (Italy).Open Access| https://doi.org/10.1155/2014/841747.

Characteristics and provenance implications of rare earth elements and Sr–Nd isotopes in PM2.5 aerosols and PM2.5 fugitive dusts from an inland city of southeastern China. Atmospheric Environment (IF4.798), Pub Date: 2020-01-01, DOI: 10.1016/j.atmosenv.2019.117069.

Chibuike, G., and Obiora, s. (2014).Heavy Metal Polluted Soils: Effect on Plants and Bioremediation Methods, *Applied and Environmental Soil Science*, vol. 2014, Article ID 752708, 12 pages, 2014. https://doi.org/10.1155/2014/752708.

Crocket, K., Hill, E., Abell, R., Johnson, C., Gary, S., Brand, T., and Hathorne, C. (2018). Rare Earth Element Distribution in the NE Atlantic: Evidence for Benthic Sources, Longevity of the Seawater Signal, and Biogeochemical Cycling. *Front. Mar. Sci.*, https://doi.org/10.3389/fmars.2018.00147.

Cynthia V. Gaonkar, Ashwini Kumar, Vishnu Murty Matta andSiby Kurian (2019). *concentrations in atmospheric particulate matter over a coastal city in the Eastern Arabian Sea*.doi.org/10.1080/10962247.2019.1680458.

Dai, Q., Li, L., Li, T., Bi, X., Zhang, Y., Wu, J., Liu, B., Gao, J., Gu, W., Yao, L., and Feng, Y. (2016). Atmospheric Signature and Potential Sources of Rare Earth Elements in Size-Resolved Particulate Matter in a Megacity of China. *Aerosol and Air Quality Research*, 16, 2085–2095, 2016 doi: 10.4209/aaqr.2016.03.0108.

Fabio, G., Silvia, b., Laura, C., David, C., marco, G., Mery, M., and Roberto, U. (2016). Metals and Rare Earth Elements in polar aerosol as specific markers of natural and anthropogenic aerosol sources areas and atmospheric transport processes. *EGU General Assembly*, id. EPSC2016-14127.

Ferrata, M., Dominik, J., Weissa, B., Strekopytovb S., Donga S., Chenc H., Najorkab, Sunc, Y., Guptaa, S., Tadad, R., and Sinha, R. (2011). Improved provenance tracing of Asian dust sources using rare earth elements and selected trace elements for palaeomonsoon studies on the eastern Tibetan Plateau. *GeochimicaetCosmochimicaActa* [*Geochemistry and CosmochemistryActa*], 75 (2011) 6374–6399.
https://scholars.unh.edu/ersc/60.

Hu, G., Wang, S., Yu, R., Zhang, Z. & Wang, X. (2019). Source Apportionment of Rare Earth Elements in PM2.5 in a Southeast Coastal City of China. *Aerosol Air Qual. Res.*, 19, 92-102. https://doi.org/10.4209/aaqr.2017.12.0559.

Jickells, T., Baker, A., and Chance, R. (2016). *Atmospheric transport of trace elements and nutrients to the oceans*. doi.org/10.1098/rsta.2015.0286.

Kolawole, T., Olatunji, O, Ajibade, O., and Oyelami, C. (2021). *Sources and Level of Rare Earth Element Contamination of Atmospheric Dust in Nigeria J Health Pollut.*, 11(30), 210611.doi: 10.5696/2156-9614-11.30.210611.

Lai, M., Shafer, M., Dibb, M., and Schauer, J. (2017). Elements and inorganic ions as source tracers in recent Greenland snow. *Atmospheric Environment*., 60.

Mahowald, N.M., Hamilton,D.S., and Mackey, K.R.M. (2018) Aerosol trace metal leaching and impacts on marine microorganisms. *Nat Commun*, 9, 2614 (2018). https://doi.org/10.1038/s41467-018-04970-7.

Malhotra, N., Hsu, H., Liang, S., Jmelou, M., Roldan, M., Lee, J., Ger, T., and Hsiao, C. (2020). An Updated Review of Toxicity Effect of the Rare Earth Elements (REEs) on Aquatic Organisms. *Animals (Basel)*. 10(9), 1663.doi: 10.3390/ani10091663.

Manisalidis, I., Stavropoulou, E., Stavropoulos, A.,and Bezirtzoglou, E. (2020). Environmental and Health Impacts of Air Pollution: A Review. *Front. Public Health*, 20 February 2020 | https://doi.org/10.3389/fpubh.2020.00014.

Marx, S. K., Lavin, K. S., Hageman, K. J., Kamber, B. S., O'Loingsigh, T., and McTainsh, G. H. (2014). Trace elements and metal pollution in aerosols at an alpine site, New

Zealand: sources, concentrations and implications. *Atmospheric Environment*, 82, 206-217.

McLennan, S.M. (2018) Lanthanide Rare Earths. In: White W. (eds) Encyclopedia of Geochemistry. *Encyclopedia of Earth Sciences Series*. Springer, Cham. https://doi.org/10.1007/978-3-319-39193-9_96-1.

Mehmood, M. (2018).Journal of Ecology & Natural Resources. A Review*J Ecol& Nat ResourRare Earth Elements*., OI: 10.23880/jenr-16000128.

Merchant, S., and Helmann, J. (2012). Elemental Economy: microbial strategies for optimizing growth in the face of nutrient limitation. *AdvMicrob Physiol.*,60, 91–210.doi: 10.1016/B978-0-12-398264-3.00002-4.

Modrzynska, J., Berthing, T., Ravn-Haren, G., Kling, K., Mortensen, A., Rasmussen, R. R., Larsen, E. H., Saber, A. T., Vogel, U., and Loeschner, K. (2018). *In vivo*-induced size transformation of cerium oxide nanoparticles in both lung and liver does not affect long-term hepatic accumulation following pulmonary exposure. *PloS one*, 13(8), e0202477. https://doi.org/10.1371/journal.pone.0202477.

Mohan, I., Goria, K., Kothari, R., and Pathania, D. (2021). Phytoremediation of Heavy Metals from the Biosphere Perspective and Solutions. *Pollutants and Water Management: Resources, Strategies and Scarcity*. https://doi.org/10.1002/9781119693635.ch5.

Moreno, T., Querol, A., Reche, M., Cusack, M., Amato, F., Pandolfi, M., Pey, J., Richard A., and Gibbon W. (2011). *Atmos. Chem. Phys. Discuss.*, 11, 14747–14776, 2011 www.atmos-chem-phys-discuss.net/11/14747/2011/doi:10.5194/acpd-11-14747-2011.

Radomskaya, V.I., Yusupov, D.V. & Pavlova, L.M. (2018). Rare-Earth Elements in the Atmospheric Precipitation of the City of Blagoveshchensk. *Geochem. Int.*, 56, 189–198 (2018). https://doi.org/10.1134/S0016702918010056.

Rai, K., Lee, S., Zhang, M., and Tsang, F. (2019). Heavy metals in food crops: Health risks, fate, mechanisms, and management.125, *Environment International*, 365-385. doi.org/10.1016/j.envint.2019.01.067.

Spickova, J., Navratil, T., Rohove, J., Mihaljevic, m., Kubinova, P., Minark, L., and Skrivan, p., (2010). *Geochemistry: Exploration, Environment, Analysis*, 10, 383-390, 10 2010, https://doi.org/10.1144/1467-7873/09-225.

Stewart, A.G. (2020). Mining is bad for health: a voyage of discovery. *Environ Geochem Health*, 42, 1153–1165. https://doi.org/10.1007/s10653-019-00367-7.

Turetta, C., Feltracco, M., Barbaro, E., Spolaor, A., Barbante, C., and Gambaro, A. (2021). A Year-Round Measurement of Water-Soluble Trace and Rare Earth Elements in Arctic Aerosol: Possible Inorganic Tracers of Specific Event. *Atmosphere*, 2021, 12(6), 694; https://doi.org/10.3390/atmos12060694.

Turetta, C., Zangrando, R., Barbaro, E., Gabrieli, J., Scalabrin, E., Zennaro, P., and Andrea Gambaro, Lincei, R. (2016). *Water-soluble trace, rare earth elements and organic compounds in Arctic aerosol*. 27, 1,95-103.

Turetta, C., Zangrando, R., and Barbaro, E (2016). Water-soluble trace, rare earth elements and organic compounds in Arctic aerosol. Rend. *Fis. Acc. Lincei*, 27, 95–103 (2016). https://doi.org/10.1007/s12210-016-0518-6.

Vlasov, D., Vasilchuk, J., Kosheleva, N., and Kasimov N. (2020). Dissolved and Suspended Forms of Metals and Metalloids in Snow Cover of Megacity: Partitioning and Deposition Rates in Western Moscow. *Atmosphere*, 2020, 11(9), 907; https://doi.org/10.3390/atmos11090907.

Wawrosl, T., Talik, E., Zelechower, M., Pastuszka, J., and Ujma, Z. (2003). Seasonal Variation in the Chemical Composition and Morphology of Aerosol Particles in the Centre of Katowice. *Pol. J. Environ. Stud.*, 2003, 12(5), 619–627.

Wiklund, J., Kirk, J., Muir, D., Gleason, A., Carrier, J. & Yang, F. (2020). Atmospheric trace metal deposition to remote Northwest Ontario, Canada: Anthropogenic fluxes and inventories from 1860 to 2010. *Science of The Total Environment.*, 749, 20 December 2020, 142276.

Xing, Y., Xu, Y., Shi, M., and Lian, Y. (2016). The impact of PM2.5 on the human respiratory system. *J Thorac Dis.*, 2016 Jan; 8(1), E69–E74.doi: 10.3978/j.issn.2072-1439.2016.01.19.

Yan Y., Zheng Q., Yu R., Hu G., Huang H., Lin C., Cui J., and Yan Y. (2020). Characteristics and provenance implications of rare earth elements and Sr–Nd isotopes in PM2.5 aerosols and PM2.5 fugitive dusts from an inland city of southeastern China. *Atmospheric Environment*, 220, 117069https://doi.org/10.1016/j.atmosenv.2019.117069.

Yannick, A., and Sejalon D., and Probst, A. (2014). Origin and distribution of rare earth elements in various lichen and moss species over the last century in France. *Science of theTotalEnvironment*, vol.487.pp.1-12. ISSN 0048-9697.

Chapter 2

Microplastics in the Atmosphere: Emerging Air Contaminants

Namita Yadav[1], Monika Vats[2], Dipanjana Chakraborty[3], Droupti Yadav[4] and Kushagra Rajendra[1,*]

[1]Amity School of Earth and Environmental Science, Amity University Haryana, Haryana, India
[2]Amity School of Applied Science, Amity University Haryna, Haryana, India
[3]Department of Geography, Sitalkuchi College, Sitalkuchi, West Bengal, India
[4]Department of Environmental Science, IBSBT, CSJM University Kanpur, Kanpur, Uttar Pradesh, India

Abstract

Plastics are miracle materials, but their extensive use is threatening the natural environment on which humanity and life critically depend, impacting aquatic and marine life, and the food chain and even detected at an unprecedented level in the atmosphere, from urban to the rural landscape and even in an indoor environment. Understanding of its atmospheric origin, life cycle, transport, deposition, impact, and interference with meteorology and climate is still at a nascent stage, possibly due to its extensive spread, diverse composition, and limited sampling and quantification methods. Microplastic forms of occurrence (fragments, shape, and nature) are aligned with the use pattern of plastics on the ground. Present study reviews contemporary knowledge on Suspended Atmospheric Microplastics (SAMP), including its physical & chemical characteristics, origin, release pathway, health exposure,

* Corresponding Author's Email: mekushagra@gmail.com.

In: Atmospheric Aerosols
Editor: Binoy K Saikia
ISBN: 979-8-88697-211-5

movement (short term and long term), and deposition. This article also reviews and suitably highlights the sampling and detection methods in an indoor and outdoor environment. Microplastic occurrence is very diverse with an abundance and deposition across locations due to local and regional meteorology. To understand its impact, this article provides an overview of health impact and explores the correlation of its short- and long-term transport flux with weather & climate. Furthermore, the summary related to the findings on atmospheric microplastic properties like diversity, abundance, colors, shape, and size will provide a ready reference. We conclude with a holistic overview of an emerging area of microplastics in the atmosphere as an air pollutant and its linkages with human/ecosystem health and meteorology.

Keywords: microplastics, aerosol, indoor environment, Suspended Atmospheric Microplastics (SAMP)

1. Introduction

Plastics are miracle materials, but their extensive use is threatening the natural environment on which humanity and life critically depend, impacting aquatic and marine life, and the food chain and even detected at an unprecedented level in the atmosphere, from urban to the rural landscape and even in the indoor environment. The term "microplastic" was conceptualized by Thompson and this term refers to the size of particles less than 5 mm (0.2 inches) (Thompson 2004). There are two main sources where microplastics are generated: primary and secondary. Primary microplastic is little particles intended for business use, like beauty care products, just as microfibers shed from apparel and different materials, like fishing nets. Secondary microplastic is particles that outcome from the breakdown of bigger plastic things, for example, water bottles. This breakdown is brought about by openness to natural variables, primarily the sun's radiation and sea waves. The issue with microplastics is that—like plastic things of any size they don't promptly separate into innocuous particles. Plastics can require hundreds or millennia to disintegrate and meanwhile, unleash destruction on the climate. On sea shores, microplastics are apparent as little diverse plastic pieces in sand. In the seas, microplastic contamination is frequently devoured by marine creatures. A portion of this natural contamination is from littering; however, much is the consequence of tempests, water overflow, and winds that convey plastic both unblemished articles and microplastics into our seas (Allen et al. 2019).

Single-use plastics, things intended to be utilized only a single time and afterward disposed of, like straw are the essential wellspring of optional plastics in the climate (Abbasi et al. 2019). Microplastics have been distinguished in marine life forms from tiny fish to whales, and surprisingly in drinking water. Alarmingly, standard water treatment can't eliminate all hints of microplastics. To additionally confound matters, microplastics in the sea can tough situations with other destructive synthetic substances prior to being ingested by marine organic entities.

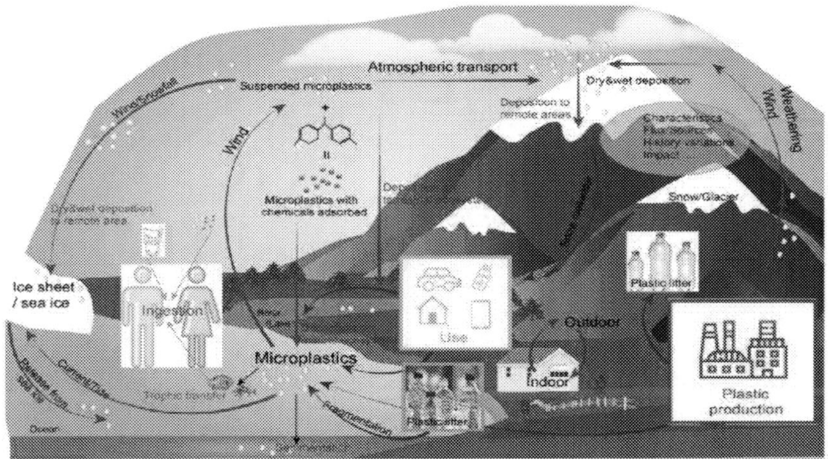

Figure 1. A model representing the microplastic in the atmosphere (Yang et al. 2021).

Several research highlighted the widespread existence of plastics in form of microplastics in the aquatic environments, streams, ponds, seashores, estuaries & terrestrial environments (Peng et al. 2017). In Three Gorges Reservoir, the concentration of microplastics varies from 1597 - 12,611 particles per m^3 & in Dongting Lake there are 900 - 2800 particles per m^3, whereas in North Atlantic Ocean the concentration of microplastics is found to be 2.46 particles per m^3 & in the Bohai Sea, in China the is found to be 0.34 particles per m^3 (Zhang et al. 2017; Lusher et al. 2014; Wang et al. 2018; Di and Wang, 2018). Reasons for microplastic pollution in aquatic systems are the release of effluent, dumping of garbage, waste-intensive sectors like the industrial sector, fishing industry, agricultural activities domestic waste & surface runoff (Liu et al. 2019c; Murphy et al. 2016). Due to the low degradation rate, microplastics might stance long-lasting fears to the ecosystems. Microplastics can be consumed mistakenly or carelessly by a diverse variety of species (Peng et al. 2017). In marine organisms (fish, shrimp, turtles, bivalves, zooplankton, and even large organisms like whales) microplastics were also observed (Devriese et al. 2015; Cole et al. 2013; Li et al. 2015; Santos et al. 2015; Lusher et al. 2015; Ferreira et al. 2016).

2. Microplastic as Health Stresser

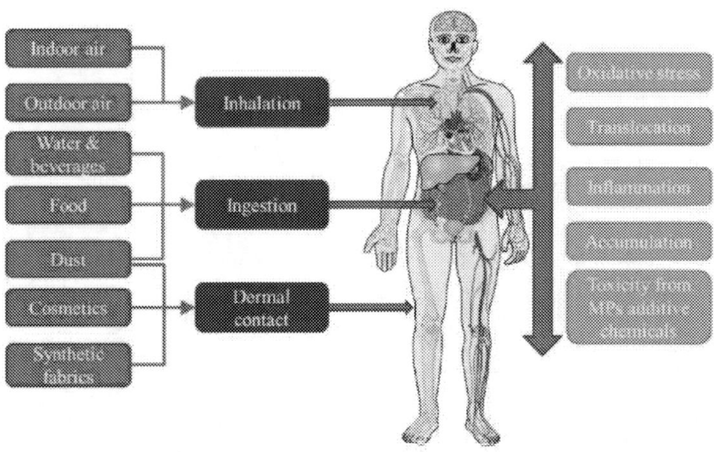

Figure 2. Exposure pattern of Microplastics and the risks associated with MPs on human health (Ageel, Harrad, and Abdallah 2022).

Airborne microplastic pollution became a new human and animal stressor as well as a threat to the healthy survival of organisms, ecological imbalance, and many more. Anthropogenic activities that generate Microplastics (MPs), Nanoplastics (NPs) and get released into the atmosphere have become ubiquitous in recent times. Due to the consumption of microplastics by human and widespread exposure, major threats (physical and physiological) can occur in their body, which includes blocking, swelling, reduction in the growth rate, increase in oxidative stress, and complications in reproduction (Alomar et al. 2017; Wright et al. 2013; Von Moos et al. 2012). Several epidemiological research shows an intense connection between air pollutants and human health especially respiratory and cardiovascular diseases (Prata 2018). Plastic fibers mixed with dust particles inhaled by people cause lung diseases that further lead to lung cancer in extreme situations. But the exposure is much more hazardous for textile workers and those working in packaging and polymer industries (Chen et al. 2020). Indoor sources of MPs are also hazardous as outdoor MPs. Fibers originating from plastics are the common source of airborne microplastics that will make up to 60 million metric tons of the world's plastic production, which results in extensive health problems like lung inflammations, infertility, and cancer because microplastics carry chemicals easily (Gasperi J et al. 2019). Contaminants like poisonous chemicals and toxic metals get absorbed by microorganisms through the microplastic present on the surface. Microplastics have larger specific surface area & stronger sorption capacities that will pose a potential threat to aquatic life (Reisser et al. 2014; Koelmans et al. 2013). In the food chain, microplastics can transfer without any difficulty. Bioaccumulation of microplastic in the food chain can pose threat to human health and life (Seltenrich, 2015; Rochman et al. 2015). That is why it has become a prime concern for environmentalists, and researchers globally in recent days.

3. Airborne Microplastics

In the past research was conducted to establish the role of microplastic as a pollutant is mostly related to terrestrial and marine ecosystems. Recently the focus swung in relation to the hazardous effect of airborne microplastic pollution (Dehghani, Moore, and Akhbarizadeh 2017). Airborne microplastic of urban, rural, and hinterlands has the potential to spread its carcinogenic effect through air, Soil & water bodies (Brahney et al. 2021). Microplastics that are 1μm to < 5mm in size, float in the air and are suspended for longer during landfall (Cai et al. 2017). They may be organic or inorganic in nature generated from plastic waste. The overuse and inefficiency in plastic waste management generate MPs in the environment. Biodegradable and non-biodegradable plastics used in the packaging industry generate MPs that enter the air during processing and at the time of recycling (Biber et al. 2019). Presence of MPs in both open and indoor air results in air quality degradation, which spreads very fast from one place to another. Several experimental findings show that these types of microplastics are present in the air in three major forms such as fibers, particles, and granules in varied densities and colors from the source region to the destination (Dris et al. 2017). The spatial and temporal variation of microplastics in the air is evident as atmospheric transportation of MPs now becomes a global issue.

After plastic gets scratched, tattered, and destructed microplastics are generated and these microplastics enter the atmosphere through different modes. For example, if we are wearing polyester clothes and that clothes get rubbed with surfaces then thousands of fibres get released into the atmosphere, this is how in daily life we produce tonnes of microplastic unknowingly. According to several studies, microplastic is everywhere from mountains to the oceans and is reaching everywhere. A research concluded that widespread occurrence of high concentration microplastics in snowfall everywhere in the world, even in remote locations such as Greenland and Svalbard where it is found 1,760 microplastics/liter, it demonstrates that microplastics can travel through the upper atmosphere & can get deposits any place in the world (Bergmann M et al. 2019).

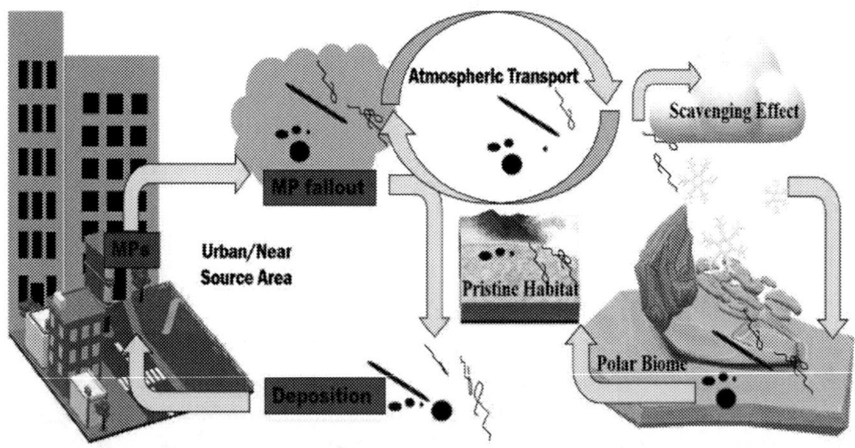

Figure 3. Representation of transportation of microplastics in the atmosphere (Sridharan et al. 2021).

The inter-linkage of source, pathway, and destination influences the trace of microplastic in the atmosphere (Zhang, 2020). The concentration of microplastic in the environment including air, aquatic, and land is influenced by atmospheric transportation (Allen et al. 2019, Zhang et al. 2019). The impact of atmospheric transportation resulted in the accumulation of suspended microplastics in ocean water. But the mechanism at the source of atmospheric transportation of microplastics is not been much explored till now. The relationship between weather and climatic elements and atmospheric transportation in diverse locations differs, which needs to be explored. It is also found that the local source of MPs is not the only reason for contamination in the air as they can be transported by air long distances (Allen et al. 2019).

General circulation of wind carried MPs to the remotest area. This is evident from the observation of microplastics on Tibetan glaciers (Zhang et al. 2019).

One of the studies on atmospheric microplastics shows that pollutants can travel more than 100km (Allen et al. 2019). This study was based on the atmospheric Hybrid Single Particle Lagrangian Integrated Trajectory Model (HYSPLIT) (Zang, 2019) to explain the mechanism of atmospheric transport. The terrestrial and oceanic sink of microplastics is highly interconnected with the atmospheric transport of pollutants. It has an influence on the retention of MPs in our environment (Horton and Dixon, 2018). MPs may produce higher toxic elements by absorbing co-existing pollutants (Sridharan et.al. 2021; Wang et al. 2021). According to Wang et al. 2021 exposed MPs is more dangerous than unexposed as they can absorb copper and tetracycline more effectively than unexposed MPs.

4. Characteristics of Microplastics

Microplastics have carbon (C) and hydrogen (H) atoms as their constituents, which were bound in a polymer chain. Further chemicals, for instance, Phthalates, Polybrominated Diphenyl Ethers (PBDEs) & Tetra Bromo Bisphenol A (TBBPA), were characteristically also existing as additives in microplastics. These chemical additives may leach out of the plastics due to environmental exposure. In the year 2006, the occurrence & abundance of Microparticles which have a size of 1.6 μm (smaller than this size) was observed for the very first time at the coastline of Singapore. Sediments from Beach & Seawater were collected from nine locations on the coastline for sampling which were - St. John Island, Pasir Ris, Semba Wang, East Coast, Changi, Sentosa Island & Kallang River. Synthetic Polymer of Microparticles was identified at sediments of the beach have PS, PP, PE, Poly Vinyl Alcohol, Nylon, and Acrylo Nitrile.

The chemical and physical properties are critically dependent on the size of the particles, shape of the particles, surface area of the particles, and crystallinity along with this chemical composition. MPs characteristics are also subject to the type of polymer, additive compounds, and the changes in surface.

Table 1. Information (characteristics, use, and production) on PP, PS, PE, PVC, PA, and PET

Plastic	Chemical formula	Commonly used for Polypropylene	Estimated degradation time (Years)	Estimated worldwide production
Poly Propylene	$(C_3H_6)_n$	Packing, clothing, and medical items	20–30	56.0 million tons
Poly Ethylene	$(C_2H_4)_n$	Packing	Up to 1000	Over 80 million Tons
Poly Styrene	$(C_8H_8)_n$	Disposable food packing	Up to 500	14.7 million tons
Poly Vinyl Chloride	$(C_2H_3Cl)_n$	Building and construction, health care, electronics, automobile	Up to 100	44.3 million tons
Poly Ethylene Terephthalate	$(C_{10}H_8O_4)_n$	Bottles, jars, containers, and packaging applications	Up to 450	30.3 million tons
Poly Amide		Textiles, automotive industry, carpets, kitchen utensils, and sportswear	-	3.4 million tons

Fotopoulou and Karapanagioti, 2019; Chamas et al. 2020; Singh and Sharma, 2008; Geyer, 2020.

4.1. Physical Properties

4.1.1. Particle Size

It was observed during the interaction between the particle and the biota that particle size plays a critical role (Montes-Burgos et al. 2010). Many studies have discovered the contrary responses of specific organisms to microparticle contact generally use uniform-sized and shaped Nano micro-beads & submicron (Cole et al. 2013; Lee et al. 2013; Besseling et al. 2014). Plastics that will experience the courses of degradation will produce a piece of the forming particles that will comprise the wide and varied range of particle size and shape supply. The separation of fragmented microparticles according to particle size is observed because of its Poly-Dispersed nature, which will increase the concentration and will decrease the distribution size (Lambert and Wagner 2016). The feeding and digestion apparatus of a particular species which came in contact with microparticles play a critical role in the absorption of microplastic in the organism (Burns 1968; Rosenkranz et al. 2009).

4.1.2. Particle Shape

Important property for the identification of particle shape is interaction between the polymeric particles and the biological system of organisms (Wright et al. 2013). According to the recent studies it was found that how the shape of the particle marks its impact on the *Hyalella azteca* (amphipod); results of this study have shown that the toxicity is high for the PolyPropylene (PP) fibers as compared to the PP beads (Au et al. 2015). In the embryos of Zebrafish, the induced toxicity in Nano sticks of Zinc oxide were found higher than the nanospheres during the evaluation of the Mortality and Hatching Inhibition (Hua et al. 2014). According to the research studies it was observed that particle with irregular shapes is capable of more readily attachment with external and internal surfaces for experience of the better effect.

4.1.3. Surface Area

Surface area is considered a critical parameter as it is inversely proportional to the particle size; so, greater effects can be induced by nanoscale particles (Van et al. 2008). Even though the surface area was not that much observed during the study of microparticles but for the primary micro-beads it was considered based upon the spherical equivalent diameter, nonetheless, for irregular-shaped secondary microparticles, this was considered as the cause of an overestimation (La Rocca et al. 2015). La Rocca and his cofounders found that the surface area of nanoscale soot particles can be estimated by using different geometrical estimates that can give the overestimation of the surface area which is sevenfold tells that a factor which is needed to be applied to get the correct estimation is the particle shape.

4.1.4. Polymer Crystallinity

Crystallinity is a significant polymer property because the translucent district comprises more arranged and firmly organized polymer chains. This influences the actual properties like thickness and porousness, which will thusly drive their hydration and expanding conduct. With time crystallinity of the Environmental Microparticles is going to transform. The better degradation of the polymer (amorphous region) is the possible cause of the total increase of crystallinity (Chen et al. 2000; Gopferich. 1996) because microparticle decreases in size. Due to this reason, crystallites will form with a difference in toxicity as compared to the original microplastic. Due to variations in crystallinity environmental microparticles may vary from their micro-bead counterparts, & this influences their other properties (physical & chemical) and will directly affect the rate of ingestion and outcomes.

4.2. Chemical Properties

4.2.1. Polymer Type and Additives

Leakage of chemicals which includes monomer residual, solvent & catalyst along with the additives like dyes and antioxidants which were used during processing cause plastic-related toxicity (Andrady, 2015; Muncke, 2009). Common monomers and additives were used in the production of certain plastic types after their toxicological profiling. Types of polymers with examples are mentioned below:

i. Plasticised Polyvinyl Chloride (PVC): It is the most hazardous plastic because it has a high content of chloride and additive & dioxins were released during the processes which include manufacturing and incineration (Rossi and Lent. 2006).
ii. Polycarbonate: Processed from Bisphenol A, which can disrupt the endocrine metabolism of organisms.
iii. Polyacrylonitriles and Acrylonitrile Butadiene Styrene: They fall under the category of carcinogenic compounds by the International Agency for Research on Cancer.
iv. PS and its copolymers: They also fall under the same category because the monomer of Styrene is an alleged carcinogen (Rossi and Lent. 2006; Lithner et al. 2011). PS is important because of its styrene oligomers which were observed in water and sediments after leaching (Kwon et al. 2015).
v. Polyurethanes and Epoxy resins: According to the classifications of their monomers within the EU classification, labeling, and packaging regulation they were also added to the previous category (Lithner et al. 2011).
vi. Polyethylene Terephthalate (PET): It has a chemical that can leach and disrupt the endocrine (Wagner and Oehlmann. 2011; Wagner and Oehlmann. 2009).

In all stages of the lifecycle, additives are released from the plastic materials into the environment based on the polymer matrix of a particular compound (Lambert et al., 2014). For example, an additive with low molecular weight will bind loosely and will move rapidly in the polymer matrix. Another example with more accuracy shows that additives that were released from various plastic materials (like flame retardants from tv, housing, and other electronic devices, nonylphenol (food packaging material), extractable PET cyclic and linear oligomers from bottles and food trays, leaching of antimony from PET water bottles and lead released from PVC pipes (Kim et al. 2006; Deng et al. 2007; Al-Malack. 2001; Fernandes et al. 2008; Kim and Lee. 2012; Shotyk and Krachler. 2007; Westerhoff et al. 2008; Keresztes et al. 2009).

Physical properties of monomers such as the diameter of the pore, the particular structure of the polymer, and the size at the molecular level of monomers & additives cause the leaching of residual monomers and additives (Gopferich. 1996). The leaching of hazardous chemicals depends on the concentration of parent plastic in the MPs along with their partition coefficient, age, and degree of degradation of the MPs. Just, for example, MP which is aged may have a high degree of crystallinity, which ultimately results in a huge decrease in leaching.

4.2.2. Surface Chemistry

With the age of MPs, their surface chemistry also changes. New functional groups are formed when OH radicals, oxygen, nitrogen oxides, and other photo-generated radicals react with the plastic surface. This reaction will lead to the Photo and oxidative degradation processes of plastic (Chandra and Rustgi. 1998). Due to this reaction, the surface of plastic starts to crack, and then it opens new surfaces which further leads to more degradation (Lambert et al. 2013). These reactive methods eventually deteriorate the plastic surface which is the root source that releases more microscopic particles during ingestion, increasing leaching of chemical & retention times will increase in the gut through the formation of particle shape more angular, which makes environmental MPs clearly more different to primary micro-beads. Additionally, interactions between MPs and organisms are affected by the change in surface chemistry (Gerritsen and Porter 1982). The surface of the plastic became more available for microorganisms as they can utilize it for their oxy-generated functional groups (Roy et al. 2008). The chemistry of MPs influences the uptake, retention, and internalization of quantum dots according to the Study with *D. magna* (Feswick et al. 2013), it was further observed that surface chemistry also plays role in cell uptake & translocation of fullerenes (Ke and Lamm. 2011). It is still unidentified that if these variations in the surface chemistry are important determinants of toxicity in environmentally accurate exposure scenarios or not.

5. Collection Methods of Airborne Microplastic

There are two types of collection methods for airborne microplastics which are Passive & active sampling methods. In the passive sampling method, the device is mostly made of a collecting funnel, terminal collection bottles & receiving tubes (Dris et al. 2015), while in the active sampling method it used to be held by pumping samplers (Dris et al. 2017; Liu et al. 2019b). The active sampling method has athe bility to efficiently decrease the time of sampling and make information available regarding the microplastics in the air mass and didn't settle down further. While passive sampling method provides information about the microplastics that fall on the surface and finally settle down (Rahman et al. 2021).

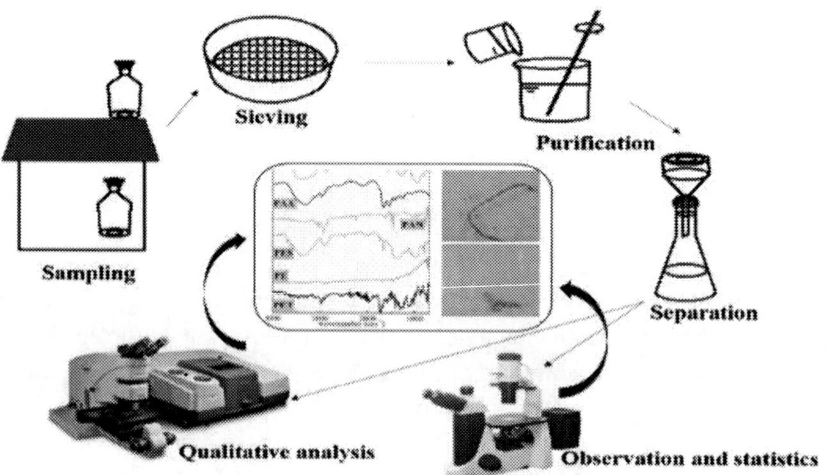

Figure 4. Flow diagram showing the steps used for the analysis of airborne microplastics (Liu et al. 2019b).

6. Analytic Methods for the Analysis of Microplastic

6.1. Microscope

In microscope sample size which is required for the analysis is down the micron range. This method of analysis quickly tells about the abundance and other physical properties of Microplastic. This method cannot determine the composition of microplastic.

6.2. Fourier Transform Infra-Red (FT-IR)

Samples with a large particle size of around >500 mm, their analysis can be done by ATR-FTIR, and particles smaller than 20 mm can be analyzed by microscope coupled FTIR. FTIR and its optimization technologies, such as micro-FTIR (μ-FTIR), Attenuated Total Reflectance FTIR (ATR-FTIR), and Focal Plane Array FTIR (FPA-FTIR) allow detection limit of microplastics up to 5–10 μm. FPA-FTIR has the ability to scan automatically the sample filter & find the spectral information speedily. It has a large database of polymers. This method cannot analyze the wet samples (Liu et al. 2019).

6.3. Raman Spectroscopy

The microscopy coupled Raman Spectroscopy (RS) is used for the analysis of particles size >1 μm. It can detect microplastics with sizes smaller than 1 mm (Levermore et al. 2020). Analysis of the wet sample & simultaneously identifies the fillers and pigments. The fast-chemical mapping can be also done by RS method.

6.4. Mass Spectrometry Analysis

In the mass spectroscopy analysis of MPs, the size of the sample is not specified. This technology does not require the pre-treatment of samples. Manual sample placement is required in pyrolysis tube, which limits its extensive application. Only one particle with a specified weight can be analyzed per run, which is another limitation. So, because there is no pre-treatment required that's why it is a fast sample analysis.

7. Looking Ahead

Due to lack of proper data and inadequate research work on MPs and its long-term exposure to human health and the environment, the estimation of long-term effects is still unquantified. A global monitoring system should develop to identify and estimate microplastic pollution through the application of standardized methods. A global database and mapping of MPs can be used for understanding the impact of MPs in the atmosphere and processes. Awareness building on Microplastics as an emerging potential pollutant in the atmosphere is highly recommended.

In the recent past, microplastics have been slowly recognized as widespread contaminants not only in aquatic/marine environments but also in the atmosphere due to their wider mobility, size, and density and it is found to have wider health implications. Although its role in interfering Earth's climate is not yet established as like for other atmospheric aerosols, its presence in the atmosphere has been widely accepted as a health stressor through aquatic and food chain routes. Plastic production has increased rapidly over the past 70 years which is associated with unregulated use and its disposal regime. It is expected that its abundance in the atmosphere will be multifold and its potential role to influence radiative forcing. There is a need for a focused attempt on monitoring microplastics in the atmosphere as like in the biological system and food chain so that their critical role as climate influencers can be assessed.

References

Abbasi, Sajjad, Behnam Keshavarzi, Farid Moore, Andrew Turner, Frank J. Kelly, Ana Oliete Dominguez, and Neemat Jaafarzadeh. 2019. "Distribution and Potential Health Impacts of Microplastics and Microrubbers in Air and Street Dusts from Asaluyeh County, Iran." Environmental Pollution 244:153–64. doi: 10.1016/ j.envpol.2018.10.039.

Ageel, Hassan Khalid, Stuart Harrad, and Mohamed Abou Elwafa Abdallah. 2022. "Occurrence, Human Exposure, and Risk of Microplastics in the Indoor Environment." Environmental Science: Processes and Impacts 24(1):17–31. doi: 10.1039/d1em00301a.

Allen, Steve, Deonie Allen, Vernon R. Phoenix, Gaël Le Roux, Pilar Durántez Jiménez, Anaëlle Simonneau, Stéphane Binet, and Didier Galop. 2019. "Atmospheric Transport and Deposition of Microplastics in a Remote Mountain Catchment." Nature Geoscience 12(5):339–44. doi: 10.1038/s41561-019-0335-5.

Alomar C, Sureda A, Capo X, Guijarro B, Tejada S, and Deudero S. 2017. Microplastic ingestion by Mullus surmuletus Linnaeus, 1758 fish and its potential for causing oxidative stress. Environ Res. 159:135-142.

Andrady AL. 2015. Plastics and Health Impacts. In: Plastics and Environmental Sustainability. John Wiley & Sons, Inc, pp 227-254.

Au SY, Bruce TF, Bridges WC, Klaine SJ. 2015. Responses of Hyalella azteca to acute and chronic microplastic exposures. Environ Toxicol Chem 34 (11):2564-2572.

Bergmann, M, Mützel, S, Primpke, S, Tekman, MB, Trachsel, J, & Gerdts, G. 2019. White and wonderful? Microplastics prevail in snow from the Alps to the Arctic. Science Advances, 5(8), eaax1157. doi: 10.1126/sciadv.aax1157.

Besseling E, Wang B, Lürling M, Koelmans AA. 2014. Nanoplastic Affects Growth of S. obliquus and Reproduction of D. magna. Environmental Science & Technology 48 (20):12336-12343. doi:10.1021/es503001d.

Biber NFA, Foggo A, Thompson RC, 2019. Characterising the deterioration of different plastics in air and seawater. Mar. Pollut. Bull. 141, 595–602. https://doi.org/10.1016/j.marpolbul.2019.02.068.

Brahney, Janice, Natalie Mahowald, Marje Prank, Gavin Cornwell, Zbigniew Klimont, Hitoshi Matsui, and Kimberly Ann Prather. 2021. "Constraining the Atmospheric Limb of the Plastic Cycle." Proceedings of the National Academy of Sciences of the United States of America 118(16):1–10. doi: 10.1073/pnas.2020719118.

Burns CW. 1968. The relationship between body size of filter-feeding cladpcera and the maximum size of particle ingested. Limnology and Oceanography 13 (4):675-678. doi:10.4319/lo.1968.13.4.0675.

Cai, Liqi, Jundong Wang, Jinping Peng, Zhi Tan, Zhiwei Zhan, Xiangling Tan, and Qiuqiang Chen. 2017. "Characteristic of Microplastics in the Atmospheric Fallout from Dongguan City, China: Preliminary Research and First Evidence." Environmental Science and Pollution Research 24(32):24928–35. doi: 10.1007/s11356-017-0116-x.

Chamas, A, Moon, H, Zheng, J, Qiu, Y, Tabassum, T, Jang, JH, Abu-Omar, M, Scott, SL, Suh, S, 2020. Degradation rates of plastics in the environment. ACS Sustain. Chem. Eng. 8, 3494–3511. https://doi.org/10.1021/acssuschemeng.9b06635.

Chandra R, Rustgi R. 1998. Biodegradable polymers. Prog Polym Sci 23 (7):1273-1335.

Chen DR, Bei JZ, Wang SG. 2000. Polycaprolactone microparticles and their biodegradation. Polymer Degradation and Stability 67 (3):455-459.

Chen G, Feng Q, Wang J, 2020. Mini-review of microplastics in the atmosphere and their risks to humans. Sci. Total Environ. 703, 135504 https://doi.org/10.1016/j.scitotenv.2019.135504.

Cole M, Lindeque P, Fileman E, Halsband C, Goodhead R, Moger J, Galloway TS 2013 Microplastic Ingestion by Zooplankton. Environmental Science & Technology 47 (12):6646-6655.

Dehghani, Sharareh, Farid Moore, and Razegheh Akhbarizadeh. 2017. "Microplastic Pollution in Deposited Urban Dust, Tehran Metropolis, Iran." Environmental Science and Pollution Research 24(25):20360–71. doi: 10.1007/s11356-017-9674-1.

Deng WJ, Zheng JS, Bi XH, Fu JM, Wong MH. 2007. Distribution of PBDEs in air particles from an electronic waste recycling site compared with Guangzhou and Hong Kong, South China. Environ Int 33 (8):1063-1069.

Di M, and Wang J. 2018. Microplastics in surface waters and sediments of the Three Gorges Reservoir, China. Sci Total Environ. 616-617:1620-1627.

Dris, R, Gasperi, J, Mirande, C, Mandin, C, Guerrouache, M, Langlois, V, & Tassin, B. 2017. A first overview of textile fibers, including microplastics, in indoor and outdoor environments. Environ Pollut 221:453–458.

Dris, R, Gasperi, J, Rocher, V, Saad, M, Renault, N, & Tassin, B. 2015. Microplastic contamination in an urban area: a case study in Greater Paris. Environ Chem 12:2015.

Dris, Rachid, Johnny Gasperi, Cécile Mirande, Corinne Mandin, Mohamed Guerrouache, Valérie Langlois, and Bruno Tassin. 2017. "A First Overview of Textile Fibers, Including Microplastics, in Indoor and Outdoor Environments." Environmental Pollution 221:453–58. doi: 10.1016/j.envpol.2016.12.013.

Fernandes AR, Rose M, Charlton C. 2008. 4-Nonylphenol (NP) in food-contact materials: Analytical methodology and occurrence. Food Additives and Contaminants 25 (3):364-372.

Feswick A, Griffitt RJ, Siebein K, Barber DS. 2013. Uptake, retention and internalization of quantum dots in Daphnia is influenced by particle surface functionalization. Aquat Toxicol 131:210-218.

Fotopoulou, KN, Karapanagioti, HK, 2019. Degradation of various plastics in the environment. Handbook of Environmental Chemistry. Springer Verlag, pp. 71–92 https://doi.org/10.1007/698_2017_11.

Gasperi, J, Wright, SL, Dris, R, Collard, F, Mandin, C, Guerrouache, M, Langlois, V, Kelly, FJ, & Tassin, B. (2018). Microplastics in air: Are we breathing it in? Current Opinion in Environmental Science & Health, 1, 1–5. https://doi.org/10.1016/j.coesh.2017.10.002.

Gerritsen J, Porter KG. 1982. The role of surface chemistry in filter feeding by zooplankton. Science 216 (4551):1225-1227.

Geyer, R, 2020. Production, use, and fate of synthetic polymers. Plastic Waste and Recycling. Elsevier, pp. 13–32 https://doi.org/10.1016/b978-0-12-817880-5.00002-5.

Gopferich A. 1996. Mechanisms of polymer degradation and erosion. Biomaterials 17 (2):103-114.

Hua J, Vijver MG, Richardson MK, Ahmad F, Peijnenburg WJ. 2014. Particle-specific toxic effects of differently shaped zinc oxide nanoparticles to zebrafish embryos (Danio rerio). Environ Toxicol Chem 33 (12):2859-2868.

Ke PC, Lamm MH. 2011. A biophysical perspective of understanding nanoparticles at large. Physical Chemistry Chemical Physics 13 (16):7273-7283Sridharan S, Kumar M, Singh L, Bolan N S, Saha M (2021) Microplastics as an emerging source of particulate air pollution: A critical review. Journal of Hazardous Materials. 418 (2021) 126245 https://doi.org/10.1016/j.jhazmat.2021.126245.

Keresztes S, Tatár E, Mihucz VG, Virág I, Majdik C, Záray G. 2009. Leaching of antimony from polyethylene terephthalate (PET) bottles into mineral water. Sci Total Environ 407 (16):4731-4735.

Kim DJ, Lee KT. 2012. Determination of monomers and oligomers in polyethylene terephthalate trays and bottles for food use by using high performance liquid chromatography-electrospray ionization-mass spectrometry. Polym Test 31 (3):490-499.

Kim YJ, Osako M, Sakai Si. 2006. Leaching characteristics of polybrominated diphenyl ethers (PBDEs) from flame-retardant plastics. Chemosphere 65 (3):506-513.

Koelmans AA, Besseling E, Wegner A, and Foekema EM. 2013. Plastic as a carrier of POPs to aquatic organisms: a model analysis. Environ Sci Technol. 47(14):7812-

7820.juveniles of the marine fish Pomatoschistus microps: Gold nanoparticles, microplastics and temperature. Aquat Toxicol. 170:89-103.

Kwon BG, Koizumi K, Chung SY, Kodera Y, Kim JO, Saido K. 2015. Global styrene oligomers monitoring as new chemical contamination from polystyrene plastic marine pollution. J Hazard Mater 300:359-367.

La Rocca A, Campbell J, Fay MW, Orhan O. 2015. Soot-in-Oil 3D Volume Reconstruction Through the Use of Electron Tomography: An Introductory Study. Tribology Letters 61 (1):1-11.

Lambert S, Sinclair CJ, Boxall ABA. 2014. Occurrence, degradation and effects of polymer-based materials in the environment. Reviews of environmental contamination and toxicology 227:1-53.

Lambert S, Sinclair CJ, Bradley EL, Boxall ABA. 2013. Environmental fate of processed natural rubber latex. Environmental Science: Processes & Impacts. 15 (7): 1359-1368.

Lambert, S, Wagner, M, 2016. Formation of microscopic particles during the degradation of different polymers. Chemosphere 161, 510–517. https://doi.org/10.1016/j.chemosphere.2016.07.042.

Lee KW, Shim WJ, Kwon OY, Kang JH. 2013. Size-dependent effects of micro polystyrene particles in the marine copepod Tigriopus japonicus. Environ Sci Technol 47 (19):11278-11283.

Levermore, Joseph M, Thomas EL, Smith, Frank J Kelly, and Stephanie L Wright. 2020. "Detection of Microplastics in Ambient Particulate Matter Using Raman Spectral Imaging and Chemometric Analysis." Analytical Chemistry 92(13):8732–40. doi: 10.1021/acs.analchem.9b05445.

Li J, Yang D, Li L, Jabeen K, and Shi H. 2015. Microplastics in commercial bivalves from China. Environ Pollut. 207:190-195.

Lithner D, Larsson A, Dave G. 2011. Environmental and health hazard ranking and assessment of plastic polymers based on chemical composition. Sci Total Environ 409 (18):3309-3324.

Liu X, Yuan W, Di M, Li Z, and Wang J. 2019c. Transfer and fate of microplastics during the conventional activated sludge process in one wastewater treatment plant of China. Chemical Engineering Journal. 362:176-182.

Liu, K, Wang, X, Fang, T, Xu, P, Zhu, L, & Li, D. 2019 Source and potential risk assessment of suspended atmospheric microplastics in Shanghai. Sci Total Environ 675:462–471.

Liu, Kai, Xiaohui Wang, Tao Fang, Pei Xu, Lixin Zhu, and Daoji Li. 2019. "Source and Potential Risk Assessment of Suspended Atmospheric Microplastics in Shanghai." Science of the Total Environment 675:462–71. doi: 10.1016/j.scitotenv.2019.04.110.

Lusher AL, Burke A, O'connor I, and Officer R. 2014. Microplastic pollution in the Northeast Atlantic Ocean: validated and opportunistic sampling. Mar Pollut Bull. 88(1-2):325-333.

Lusher AL, Hernandez-Milian G, O'brien J, Berrow S, O'connor I, and Officer R. 2015. Microplastic and macroplastic ingestion by a deep diving, oceanic cetacean: the True's beaked whale Mesoplodon mirus. Environ Pollut. 199:185-191.

Muncke J. 2009. Exposure to endocrine disrupting compounds via the food chain: Is packaging a relevant source? Sci Total Environ 407 (16):4549-4559.

Peng, G, Zhu, B, Yang, D, Su, L, Shi, H, Li, D, 2017. Microplastics in sediments of the Changjiang Estuary, China. Environ. Pollut. 225, 283–290. https://doi.org/10.1016/j.envpol.2016.12.064.

Prata, Joana Correia. 2018. "Airborne Microplastics: Consequences to Human Health?" Environmental Pollution 234:115–26. doi: 10.1016/j.envpol.2017.11.043.

Rahman, Luna, Gary Mallach, Ryan Kulka, and Sabina Halappanavar. 2021. "Microplastics and Nanoplastics Science: Collecting and Characterizing Airborne Microplastics in Fine Particulate Matter." Nanotoxicology 15(9):1253–78. doi: 10.1080/17435390.2021.2018065.

Reisser J, Shaw J, Hallegraeff G, Proietti M, Barnes DK, Thums M, Wilcox C, Hardesty BD, and Pattiaratchi C. 2014. Millimeter-sized marine plastics: a new pelagic habitat for microorganisms and invertebrates. PLoS One. 9(6):e100289.

Rochman CM, Tahir A, Williams SL, Baxa DV, Lam R, Miller JT, Teh FC, Werorilangi S, and Teh SJ. 2015. Anthropogenic debris in seafood: Plastic debris and fibers from textiles in fish and bivalves sold for human consumption. Sci Rep. 5:14340.

Rosenkranz P, Chaudhry Q, Stone V, Fernandes TF. 2009. A comparison of nanoparticle and fine particle uptake by Daphnia magna. Environ Toxicol Chem 28 (10):2142-2149.

Rossi M, Lent T. 2006. Creating Safe & Healthy Spaces: Selecting Materials that Support Healing. In: Designing the 21st century hospital environmental leadership for healthier patients and facilities. https://www.healthdesign.org/chd/research/designing-21st-century-hospital-environmental-leadership-healthier-patients-and-familli.

Roy PK, Titus S, Surekha P, Tulsi E, Deshmukh C, Rajagopal C. 2008. Degradation of abiotically aged LDPE films containing pro-oxidant by bacterial consortium. Polymer Degradation and Stability 93 (10):1917-1922.

Santos RG, Andrades R, Boldrini MA, and Martins AS. 2015. Debris ingestion by juvenile marine turtles: an underestimated problem. Mar Pollut Bull. 93(1-2):37-43.

Seltenrich N. 2015. New link in the food chain? Marine plastic pollution and seafood safety. Environ Health Perspect. 123(2):A34-41.

Shotyk W, Krachler M. 2007. Contamination of Bottled Waters with Antimony Leaching from Polyethylene Terephthalate (PET) Increases upon Storage. Environmental Science & Technology 41 (5):1560-1563.

Singh, B, Sharma, N, 2008. Mechanistic implications of plastic degradation. Polym. Degrad. Stab. https://doi.org/10.1016/j.polymdegradstab. 2007.11.008.

Thompson, RC, Olson, Y, Mitchell, RP, Davis, A, Rowland, SJ, John, AWG, McGonigle, D, Russell, AE, 2004. Lost at sea: where is all the plastic? Science 304, 838. https://doi.org/10.1126/science.1094559.

Van Hoecke K, De Schamphelaere KA, Van der Meeren P, Lucas S, Janssen CR. 2008. Ecotoxicity of silica nanoparticles to the green alga Pseudokirchneriella subcapitata: importance of surface area. Environ Toxicol Chem 27 (9):1948-1957.

Von Moos N, Burkhardt-Holm P, and Kohler A. 2012. Uptake and effects of microplastics on cells and tissue of the blue mussel Mytilus edulis L. after an experimental exposure. Environ Sci Technol. 46(20):11327-11335.

Wagner M, Oehlmann J. 2009. Endocrine disruptors in bottled mineral water: total estrogenic burden and migration from plastic bottles. Environ Sci Pollut Res Int 16 (3):278-286.

Wagner M, Oehlmann J. 2011. Endocrine disruptors in bottled mineral water: Estrogenic activity in the E-Screen. The Journal of Steroid Biochemistry and Molecular Biology 127 (1–2):128-135.

Walczyk D, Hole P, Smith J, Lynch I, Dawson K. 2010. Characterisation of nanoparticle size and state prior to nanotoxicological studies. Journal of Nanoparticle Research 12 (1):47-53.

Wang W, Yuan W, Chen Y, and Wang J. 2018. Microplastics in surface waters of Dongting Lake and Hong Lake, China. Sci Total Environ. 633:539-545.

Wang Y, Wang X, Li Y, Liu Y, Xia S, Zhao J, 2021. Effects of exposure of polyethylene microplastics to air, water and soil on their adsorption behaviors for copper and tetracycline. Chem. Eng. J. 404, 126412. https://doi.org/10.1016/j. cej.2020.126412.

Westerhoff P, Prapaipong P, Shock E, Hillaireau A. 2008. Antimony leaching from polyethylene terephthalate (PET) plastic used for bottled drinking water. Water Res 42 (3):551-556.

Wright, SL, Ulke, J, Font, A, Chan, KLA, & Kelly, FJ. (2020). Atmospheric microplastic deposition in an urban environment and an evaluation of transport. Environment International, 136, 105411. https://doi.org/10.1016/j.envint.2019.105411.

Yang, Ling, Yulan Zhang, Shichang Kang, Zhaoqing Wang, and Chenxi Wu. 2021. "Microplastics in Soil: A Review on Methods, Occurrence, Sources, and Potential Risk." Science of the Total Environment 780:146546. doi: 10.1016/ j.scitotenv.2021.146546.

Zhang W, Zhang S, Wang J, Wang Y, Mu J, Wang P, Lin X, and Ma D. 2017. Microplastic pollution in the surface waters of the Bohai Sea, China. Environ Pollut. 231(Pt 1):541-548.

Chapter 3

Organic Compounds in Atmospheric Aerosols

John Tsado Mathew[1,*], Charles Oluwaseun Adetunji[2], Abel Inobeme[3], Musah Monday[1], Elijah Yanda Shaba[4], Yakubu Azeh[1], Abulude O. Francis[5] and Amos Mamman[1]

[1]Department of Chemistry, Ibrahim Badamasi Babangida University Lapai, Niger State, Nigeria
[2]Applied Microbiology, Biotechnology and Nanotechnology Laboratory, Department of Microbiology, Edo University Uzairue, Auchi, Edo State, Nigeria
[3]Department of Chemistry, Edo University Uzairue, Auchi, Edo State, Nigeria
[4]Department of Chemistry, Federal University of Technology Minna, Nigeria
[5]Science and Education Development Institute, Akure, Ondo State, Nigeria

Abstract

Organic compounds in the atmospheric aerosols and suspended particulate matter (PM) are a vital component of the atmosphere. Most of these compounds are volatile which affects their transport and distribution in the atmosphere. The presence of these compounds has become an issue of serious concern in recent times due to their toxic impact on human health and the environment at large, and their indispensable role as a driving force for various environmental issues. Polyaromatic hydrocarbons, polychlorinated compounds and some other components have been proven to be mutagenic and carcinogenic. Organic aerosols have also been reported to significantly affect climate and visibility. This chapter critically reviews organic compounds in the atmospheric aerosol. It discusses the occurrence and sources of these compounds, classes of organic compounds in aerosols, their impact on

* Corresponding Author's Email: johntsadom@gmail.com; jmathew@ibbu.edu.ng.

In: Atmospheric Aerosols
Editor: Binoy K Saikia
ISBN: 979-8-88697-211-5

the environment and human health, recent reports on their distribution and concentrations in atmospheric aerosols. An attempt is also made at highlighting the future trends in this regard.

Keywords: organic compounds, atmospheric aerosols, particulate matter

Introduction

Organic compounds make up a considerable portion (20-80%) of the bulk of fine particulate matter in atmospheric aerosols (Tefera et al., 2021). Organic molecules are obtained through various ways, including the combustion of vegetation along with organic matter as well as photo-oxidation mechanisms in the atmosphere. Major organic compounds in atmospheric aerosol include polyaromatic compounds, chlorinated compounds, aliphatic compounds, amongst others produced mainly from anthropogenic sources even though a few are of biogenic origin (So et al., 2019).

Aerosols in the atmosphere are known to have a significant impact on climate. The many indirect and direct impacts of aerosol particles mostly on Earth's radiative balance and ecosystems have received a lot of attention. Numerous research has been carried out on certain effects, such as the absorption and scattering of solar radiation, adjustments of atmospheric biochemical processes, changes in cloud properties, and the involvement of aerosol particles in providing essential minerals to ecological systems, all of which are encapsulated inside the International Council on Climatic Change reports. Climate factors, in turn, impact mechanisms that regulate atmospheric aerosol distributions, such as transport, emissions, deposition and transformation of aerosol particles, in addition to the impact of aerosols on climate (Tegen, and Schepanski, 2018).

Atmospheric aerosol, particulate matter suspended in the air we breathe, exerts a strong impact on our health and the environment. Controlling the amount of particulate matter in the air is difficult, as there are many ways particles can be formed by both natural and anthropogenic processes. A substantial portion of atmospheric aerosol is organic, and this organic matter is exceedingly complex on a molecular scale, encompassing hundreds to thousands of individual compounds that distribute between the gas and particle phases. Because of this complexity, no single analytical technique is sufficient. However, mass spectrometry plays a crucial role owing to its combination of high sensitivity and molecular specificity. Both offline and online approaches are covered, and molecular measurements with them are discussed in the context of identifying sources and elucidating the underlying chemical mechanisms of particle formation. There is an ongoing need to improve existing techniques and develop new ones if we are to further advance our knowledge of how to mitigate the unwanted health and environmental impacts of particles (Johnston and Kerecman, 2019).

Organic compounds constitute up to a large portion of the suspended particulate mass in different parts of the earth. These types of particulate matter are essential in a variety of environmental as well as geophysical challenges, from local difficulties to world problems (for example, climate change). However, due to the complexity of organic chemistry and also the wide range of physical properties associated including both artificial and natural organic particles, sampling and acquiring chemical information on such substances is extremely difficult. As a result of these roadblocks, we have an imperfect depiction of a possibly important aspect of atmospheric chemistry, as well as limited knowledge of the aerosol's environmental as well as geophysical effects. Given the insufficiency of quantified molecular techniques, the goal of this work is to provide a foundation for determining whatever information is required, rather than to quantify the role of organic aerosols in ecological problems (Jacobson et al., 2000; Rushdi et al., 2014).

Forested areas of countries such as Egypt, Amazonian regions, IGP, China and Europe are an ideal habitat to research secondary organic aerosols because of complex atmospheric properties, which are frequently within photochemical circumstances. Investigate the content of aerosols in situ, notably in photo-oxidation processes of organic carbon volatile compounds, while accounting for natural and anthropogenic sources, and also to understand how some of these molecules occur in the aerosols. The secondary organics detected are di-, oxo-and monocarboxylic acids, alkene derivatives, thia and aza arenes, oxy-aromatics, and several terpene photo oxidation products. This in situ experiment permitted the existence of secondary components to be confirmed, which had previously only been researched in synthetic experimental conditions (Alves and Pio, 2005).

Occurrence and Sources of Organic Compounds in Atmospheric Aerosols

Despite significant advancements in the past few years, predictive and quantitative knowledge of atmospheric aerosol chemical composition, sources, environmental consequences and transformation processes remain a key research problem in atmospheric science. This overview starts with a historical context on scientific problems about atmospheric aerosols from over decades, then moves on to a characterization of their distribution, transformation processes, sources, and physical and chemical characteristics since they are now recognized (Calvo et al., 2013).

Winter/autumn could see a rise in monocarboxylic acids, polycyclic aromatic hydrocarbons (PAH), bisphenol A and biomass burning tracers whereas summer had seen an increase in dicarboxylic-, BSOA tracers as well as hydroxycarboxylic acids. Diverse sources and production pathways for heterocyclic acids have been identified, with phthalic, benzoic, and trimellitic acids peaking in the summer and isophthalic, p-toluic, among terephthalic acids peaking in the winter/autumn. The carcinogenic, carcinogenic power and mutagenic activities of Benzo[a]pyrene-equivalents were computed and showed substantial ($p > 0.05$) improvements over the wintertime (Kanellopoulos et al., 2021).

In the study of PM2.5, which was a case study of Jinan, a generally polluted city in the North China Plain, in the winter to explore the chemical compositions, origins, and development mechanisms of organic aerosols (OAs) (NCP). PM2.5, organic components (such as fatty acids, sugars, oxygenated-PAHs (OPAHs) and polycyclic aromatic hydrocarbons (PAHs)) and carbonaceous species were 1.8–2.7 times greater in hazy episodes compared to clear episodes. Furthermore, the CWT data revealed that even during haze periods, the majority of OAs came from the surrounding as well as local areas (Li et al., 2021).

Anthropogenic activities are responsible for 60 to 80 percent of sulphur emissions. The huge quantity of pollutants emitted in various phases of industrial applications, as well as the wide diversity of contaminants, characterise industrial contamination. This kind of contaminant produced is mostly determined by the manufacturing process, technologies, or raw materials employed. Several industries that produce bricks, cement, ceramics, mining, quarrying and foundries, all emit considerable amounts of primary aerosols, whether during the manipulation or production and transportation of the raw resources used (Manisalidis et al., 2019).

Steelworks are key point-source emitters of metallic pollutants, according to the study, which was based on an investigation of an incident of pollutants from industry plumes. Because metal-rich particles were detected internally mixed with marine and/or continental chemicals, the authors stressed the importance of coagulation procedures among industrial particles and particles from other sources. Energy generation from fossil fuels is a significant source of gases that function as secondary aerosol precursors. Primary particles are created by coal waste products including sulphur, clay, chlorides, metals, and carbonates primarily mercury, as well as unburned char or coal in power plants (Csavina et al., 2011).

Emission of sea spray, discharge of soil and mineral dust (rock debris) and emission of biomass burning smoke, biogenic aerosols, and injection of volcanic material to tropospheric heights through severe eruptions are all major ecological surface primary sources of aerosol particles. Space, in the form of cosmic aerosols, contributes a tiny amount to total atmospheric aerosol load distribution, although these fine particles are thought to have only a little impact on the aerosol properties of high-altitude atmospheric regions, where the intensity of particles is always really low. As a result, cosmic rays have little effect on the low stratosphere's air characteristics of the human-environmental condition mostly in the troposphere. These morphological characteristics, optical properties, chemical composition, as well as deposition processes of the many kinds of particles generated at the surface of the planet are all distinguished through well-diversified chemical composition, deposition patterns and morphological features (Viana, et al., 2014).

Aerosols can be anthropogenic or natural, dependent upon their source. Its primary components of anthropogenic aerosols mostly in the atmosphere are industrial and urban areas, and this can as a result of traffic (road surface abrasion, exhaust emissions, tyre and brake wear, particle resuspension from pavements motorways), various industrial operations (emissions from oil refineries, power plants, and mining), construction (soil movement, excavations, and demolitions), as well as emissions from residences in this context (food cooking and heating). Biomass emissions and burning from different farming operations, on the other hand, are the principal producers of aerosols in rural regions. Oceans and seas, soil, deserts, vegetation, volcanoes, lightning and wildfires are all-natural producers of aerosols. Particles from a broad variety of sources have significantly varied chemical components, which are usually connected to their source (Penkała et al., 2018).

Secondary organic aerosol influences the burden of the particles in the atmosphere, affecting climate and air quality. Biologically active volatile organic compounds released by plants, including terpenoids, are part of the key secondary organic aerosol precursors, having isoprene leading biogenic volatile organic compound global emissions. Furthermore, as related to other terpenoids, its particle mass-produced by isoprene oxidation is rather small. Researchers show that in combinations of air vapours, isoprene, methane as well as carbon monoxide may all reduce the instantaneous mass and total mass yield produced from monoterpenes Isoprene' scavenges' hydroxyl radicals, inhibiting their interaction through monoterpenes, and to ensuing isoprene peroxy radicals scavenger highly oxygenated monoterpene compounds, according to research. The yield of low-volatility substances which might normally produce secondary organic aerosol is reduced as a result of all these effects. As a result, highly reactive chemicals that create just a little quantity of aerosol aren't always net producers of secondary organic aerosol mass, and their oxidation in combinations of atmospheric vapours might reduce both the number and mass of secondary organic aerosol particles (McFiggans et al. 2019).

Major Classes of Organic Compounds Present in Atmospheric Aerosols

Organic aerosols make up a major portion of fine-mode aerosols in the atmosphere (with *particle* diameters in the *range* of 10^{-9} to 10^{-4} m). Models of secondary organic aerosol (SOA) production and aging presume that SOA is made up of liquid particles having rapid adequate condensed phase diffusion coefficient to keep the gas phase in balance (O'Brien et al., 2014).

In the same vein, Heald, & Kroll, (2020) examines available knowledge about Organic Aerosol (OA) that is important for global climate studies and identify crucial gaps that must be filled to decrease the associated assumptions. Most of the parts needed to describe OA in a global climate model are sketched out, with such a specific focus on SOA: The main carbonaceous aerosol emission estimates, as well as SOA precursor gas emission estimations, are summarized. The most recent knowledge of condensable organic material chemical production and change is presented. The hygroscopicity of OA is reported, as well as observations of optical characteristics of organic aerosol particles. The mechanics of OA contacts between clouds, as well as parameterizations for wet and dry removal processes in global models, are discussed. This data is compiled to offer a continuous examination of the movement from the released material to the atmosphere, up to the point where the created organic aerosol has a climatic influence. At each phase of the process, the sources of uncertainty are indicated as regions that need more research.

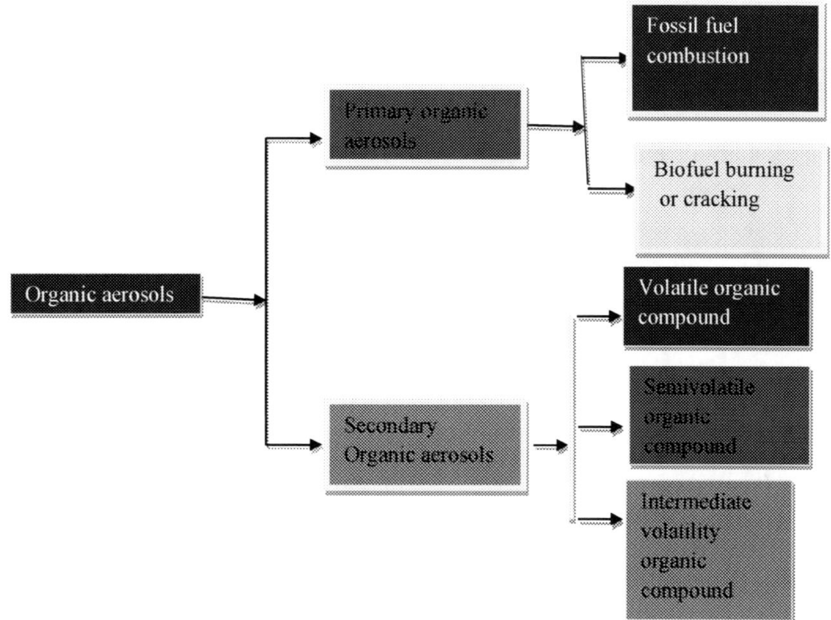

Figure 1. Major classes of organic aerosols.

However, Huanga et al. (2021) discovered secondary organic aerosol (SOA), secondary inorganic aerosol (SIA) and Primary organic aerosol (POA) can all be found in a single atmospheric particle (SIA). Research on the quantity but also kinds of phases available in such complicated multicomponent particles is essential to forecast their involvement in climate and air quality. On the other hand, SIA+POA+SOA particles can include three different liquid phases: a higher-polarity organic-rich phase, an aqueous inorganic-rich phase as well as a low-polarity organic-rich phase, as seen here. Based on their findings, three liquid phases could persist inside the same particle across a broad humidity levels scale whenever the SOA's elemental oxygen-to-carbon (O: C) ratio is much less than 0.8. When the SOA's O: C ratio is larger than 0.8, however, three phases do not emerge. It also shows that the existence of three liquid phases in these kinds of particles affects overall equilibration timing also with nearby gas-phase utilizing kinetic and thermodynamic simulations. The capacity of these particles to function as nuclei for liquid cloud droplets, their reactivity and the process of SOA creation and development in the atmosphere will all be influenced by three stages.

Marine aerosol, which contains both inorganic and organic constituents of secondary and primary origin, is among the most significant natural aerosol processes on a worldwide scale. The current research compares novel secondary and primary organic marine aerosol data obtained during the EU project Marine Aerosol Production (MAP) with some of those reported in the previous research. Following prior studies, marine aerosol samples taken at the coastal location of Mace Head, Ireland, indicate a chemical properties pattern that is impacted mostly through the oceanic biological activity cycle (Rinaldi et al., 2010).

Physicochemical Properties of Organic Aerosols

The study of Hays, (2014), stated that the chemical makeup of aerosols is complicated. Several organic molecules, black carbon and refractory brown, cations, heavy metals, salts, anions, and other inorganic phases could be found in ignition aerosols. Semivolatile component divides among the liquid, gas, as well as particle phases in aerosol organic compounds, a process governed by either interfacial mass or dynamic equilibrium transfer concerns. Photo-oxidation of semivolatile and volatile organic materials in the atmospheric could result in the formation of growing particles or particle nuclei through condensation. This heterogeneous particle nanostructure and morphology is produced by the different chemistry along with the processing of anthropogenic emissions aerosols, that impacts to an aerodynamic nature, optical characteristics, and hence the destiny and movement of aerosols. The complicated chemical composition of burning aerosols necessitates the application of several advanced scientific measuring methods. The study focuses on the measures that have emerged in the combustion and aerosol research community for accurately evaluating the chemical and physical characteristics of anthropogenic particles.

Because aerosol particles have such limited air lives, their chemical composition can fluctuate greatly depending on local sources and sinks. Submicron aerosol is made up of organic compounds, inorganic compounds, mineral species and elemental carbon. The bulk of nitrate, sulphate, elemental carbon and ammonium in today's atmosphere come through anthropogenic emissions; organic molecules come through both natural and anthropogenic sources (Allen et al., 2019).

Nitrate, ammonium as well as sulphate are the three frequent inorganic aerosol components. Ammonia (NH_4) is launched as a gas through both anthropogenic and natural sources, with agricultural practices seeming to be the most prevalent. Once before in the environment, it reacts using different acids to generate ammonium salts. Sulphate has been most typically created when sulphur dioxide, which is produced by anthropogenic activity, the volcanoes or sea, is oxidized. Sulphuric acid (H_2SO_4) could persist as either a particle mostly in the atmosphere, although in the presence of ammonia, it neutralizes to create salts such as ammonium bisulphate (NH_4HSO_4) and ammonium sulphate ($(NH_4)_2SO_4$. Nitrate is generated in the atmosphere when NO_2 is neutralized. Particulate ammonium nitrate (NH_4NO_3) is formed when there is an accumulation of ammonia in the environment once sulphate has now been neutralized (Zhu et al., 2015; Farah, 2018).

Tropospheric aerosols have an impact on global temperature, air quality and atmospheric compounds. According to current assessments, organic chemicals in this haze account for about half of the total aerosol fine mass concentration worldwide. Although, the well-constrained mechanisms that produce sulfate or nitrate aerosol, oxidation of volatile organics in the atmosphere could produce hundreds of persistent chemicals in the aerosol phase. Modeling the organic aerosol's influence on global warming requires the development of such an amenable system that takes into account the physical and chemical development of the organic aerosol. Researchers explain how to combine a three-dimensional coordinate system described through heteroatom mass, double bond equivalents (D.B.E.) and molecular weight with high-resolution molecular mass data. The use of spectrometry to describe important features of organic aerosols is a strong tool. The technique is simple in principle yet complicated enough to work with quantitative structure-property relationships (QSPRs), which are utilized to estimate physical and chemical factors that regulate aerosol behavior (Yiyi et al., 2012).

If methane or ethane condensate clouds exist, they are patchy, thin, and ephemeral. The pole shadow appears to be connected with, but not contained in, stratospheric clouds of compressed hydrocarbons and nitriles, implying that abundances might change by the season. Crystallizing condensate particles again from the stratosphere is most likely crystallizing centers for the creation and fast expansion of methane ice particles in the troposphere, where even the gas phase is extremely fully saturated. Some hailstones have a fallout period of 2 hours or less once generated. Melting and the possible breakup of methane droplets should happen at a depth of 12 km or less (McKay et al., 2001).

Transport Mechanisms of Organic Compounds in Atmospheric Aerosols

Aerosols in the atmosphere include a large amount of water in the form of thin water films and bulk water. Organic molecules with low vapor pressures and limited solubility give adsorptive surfaces and very large surface areas. They likewise provide very active surface area for the oxidation process in the gaseous environment utilizing singlet oxygen, ozone species as well as hydroxyl. As a result of the interaction at the air-water interface of cloud/fog droplets and in the thin water layer organic molecules and atmospheric aerosols are easily converted (Kalliat et al., 2015).

For responses to the questions "What is there and also how did it get there," atmospheric aerosol research has depended primarily on measurements. Furthermore, researchers have depended on the principles of aerosol research for the assessment of atmospheric particle properties, in addition to recording the composition of atmospheric aerosols. Even by the 1990s, the prominent components of the scientific research that apply to atmospheric particles had become well documented. Theoretical aspects of aerosol scientists that piqued my interest were chemical and mechanics mechanisms (Hidy, 2019).

Secondary organic aerosols are the type of aerosol that forms in the atmosphere when low vapor pressure organic contaminants condense (SOA). SOA is thought to provide an impact mostly on affecting climate change, the atmosphere's radiation budget, public health as well as visibility. Previous research has shown that SOA may considerably influence the fine particle load in the atmosphere in metropolitan areas, particularly when strong smog events (Song et al., 2005).

Recent field data reveals that organic compounds constitute a large constituent of atmospheric aerosols, which supports this concept. A "reversed micelle" with an aqueous core encased in an inert, hydrophobic organic monolayer is suggested as the model organic aerosol. Surfactants of natural material cover the aerosol particles with organic components. They suggest a chemical mechanism in which air radicals interact through the organic surface layer to metabolize it. When an organic aerosol is subjected to an oxidizing environment, an optically active hydrophilic layer and an inert hydrophobic coating transform into a reactive one. As a result, water accumulation could cause packaged organic aerosols to develop and form cloud deposition nuclei, affecting atmospheric radiative transmission. The chromophores that remained on the surface of the aerosol following chemical conversion might alter radiant transmission significantly. The chemical model generates predictions that may be verified through observation. A curve of percentage organic content as a function of particle diameter, for example, indicates that such upper tropospheric aerosol would include a large proportion of organic substances. Organic aerosol processing in the atmosphere will result in the release of tiny organic pieces into the troposphere, which will then play a role (Ellison et al., 1999; Dobson et al., 2000; Su et al., 2020).

Impact of Organic Contaminants on Human Health and the Environment

Combustion of fossil fuels, transportation, chemical industries, chlorination water purification and pesticides utilized in agriculture are only a few of the forms of organic contaminants released into the environment. The presence of these compounds presents a serious threat to human health, the environment, and other beings (Patel et al., 2020).

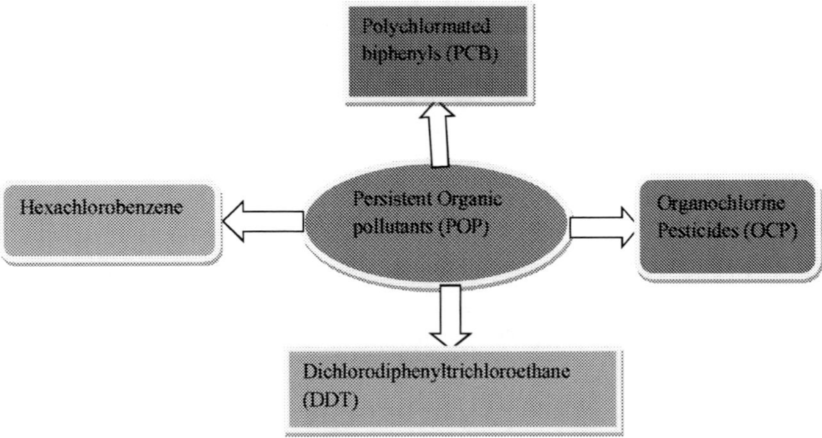

Figure 2. The major class of toxic organic compounds.

Persistent organic pollutants (POPs) are a type of harmful organic chemical that is released into the environment as a result of various human activities. Due to various durability, bioaccumulation, toxicity, and vulnerability, these chemical compounds have become a significant source of worry. POPs include compounds including polychlorinated biphenyls (PCBs), dichlorodiphenyltrichloroethane (DDT), (Figure 2) as well as by-products including furans and dioxins, among others. Continuous releases of hazardous chemicals into the water, soil, and sediments have far-reaching consequences for the whole planet (Kudom, 2019).

PCBs, organochlorine insecticides, hexachlorobenzene, DDT, and other industrial by-products such as dibenzofurans, polychlorinated dibenzo-p-dioxins as well as polynuclear aromatic hydrocarbons, are all examples of POPs. Synthetic, lipophilic, persistent, extremely poisonous organic contaminants that accumulate in the food system. POPs bioaccumulate in adipose tissue, raising the likelihood of negative health impacts in children (Thakur and Deepak, (2020).

Recent Reports on Concentrations and Distributions of Various Organic Compounds in Atmospheric Aerosols

The current state of knowledge on the source, composition, impacts and transformations of water-soluble organic aerosol (OA) has still been restricted to outdoor situations. The research, which was influenced through investigations on environmental OA, analyzes but also highlights topics connected to domestic water-soluble OA and its implications on human wellbeing, laying the groundwork for future research on the subject. The following are the three primary subjects that are discussed: the present condition on WSOM from the indoor particles in the air, including the major opportunities and challenges for cytotoxicity evaluation and chemical characterization; what is known about the mass contribution, origin as well as health effects of WSOM in outdoor air particles; and why the aerosol WSOM should have been suggested for future indoor air quality research findings. While difficult, research on the WSOM component in particles of the air is critical for a complete understanding of its fate, origin, long-term dangers inside as well as toxicity (Duarte and Duarte, 2021; Haque et al., 2019).

$PM_{2.5}$, organic components (such as fatty acids, sugars, oxygenated-PAHs (OPAHs) and polycyclic aromatic hydrocarbons (PAHs) and carbonaceous species were 1.8–2.7 times greater in hazy episodes than in clear episodes (Jelena et al., 2021). One of the most prevalent saccharides has been levoglucosan, which showed substantial associations between carbonaceous species throughout the sample period, indicating that biomass burning has a major impact on carbonaceous species amounts in PM2.5 in Jinan throughout the winter. During the whole sample period, the positive matrix factorization (PMF) model revealed that biomass burning had been the primary source of OAs. Research outcomes of the concentration-weight trajectory (CWT) and the potential source contribution function (PSCF) show that Shandong Province and the Beijing-Tianjin-Hebei (BTH) are contributed significantly to PM2.5 levels in Jinan over the winter. Furthermore, the CWT data revealed that during hazy periods, the majority of OAs came from the local and adjacent areas (Li et al., 2021).

Conclusion

Considering major recent scientific advances and efforts, significant knowledge gaps of atmospheric organic aerosols remain, impeding attempts to comprehend, mitigate and model environmental concerns including aerosol generation both in more pristine locations and polluted urban. During the last century, the experimental toolbox accessible to scientists studying atmospheric organic substances had also grown significantly, offering new insights into time resolution, speciation and identification of semi volatile and reactive molecules at trace levels. It has opened up previously unimagined possibilities while simultaneously presenting modern research obstacles. The involvement of epoxides throughout the aerosol formation, particularly from isoprene, its significance of common oxidizing, responsive organic through air-surface mechanisms (regardless of whether aerosol or atmosphere-biosphere exchanges), as well as the extensiveness of interactions between biogenic and anthropogenic emissions and indeed the subsequent influence on atmospheric organic aerosols are just a few instances of ground-breaking research. For carbonaceous aerosols, improved monitoring approaches are required, particularly for quite a better knowledge of the aging processes and chemical composition of organic aerosols. When collecting as well as analyzing organic aerosol chemicals, special care must be taken to remove errors. Because the difference between organic and black carbon is not always evident, measurements must be supported with a characterization of a quantity being measured. It is necessary to attempt to avoid discrepancies in carbonaceous aerosol readings.

Future studies should focus on basic photochemical mechanisms, enhanced chemical deposition, speciation, nitrogen budget closure, and reactive carbon and seasonal and geographical variety in chemical systems, building on recent breakthroughs in the scientific evidence of atmospheric organic aerosols. This information should have been tailored to individual social concerns such as climate change and the effects of air pollution on human health, as well as appropriate answers such as urban growth, land use modifications, and the creation of new power generation. As a result, societal reactions will have significant feedback effects on atmospheric aerosols, which our society should work to comprehend and anticipate. Particularly tiny organic molecules might experience photochemical degradation, resulting in the creation of 10 to 1000 of those first-generation devices, which will subsequently experience additional transformation and oxidation. As a result, atmospheric researchers are unlikely to be able to detect most VOC oxygen-free radicals mostly in the environment. Therefore, scientists should aim towards diversification to fully comprehend the complexity of detectable constituents to the point from which we can grasp AOAs at the operational level. A deeper knowledge of AOAs at the borders between macro and micro scales is also required. Gas-particle interactions, including those among the biosphere and the atmosphere, hydrosphere and cryosphere, are all examples. Mechanisms in the aqueous medium, along with interactions among organic and inorganic constituents of aerosols, need to be investigated more since they are likely to disclose previously unknown mechanisms.

References

Allen S. A. A., Ree A. G. and Ayodeji, S. A. M. (2019). Secondary inorganic aerosols: impacts on the global climate system and human health. *Biodiversity of International Journal,* 3(6):249-259. doi: 10.15406/bij.2019.03.00152.

Alves, C. A. and Pio, C. A. (2005). *Secondary organic compounds in atmospheric aerosols: speciation and formation mechanisms. Journal of the Brazilian Chemical Society, 16(5), 1017–1029.* doi:10.1590/S0103-50532005000600020.

Calvo, A. I., Alves, C., Castro, A., Pont, V., Vicente, A. M. and Fraile, R. (2013). *Research on aerosol sources and chemical composition: Past, current and emerging issues. Atmospheric Research, 120-121, 1–28.* doi:10.1016/j.atmosres.2012.09.021.

Csavina, J., Landázuri, A., Wonaschütz, A., Rine K., Rheinheimer, P., Brian, B., Conant W., Sáez E. A. and Betterton, E. A. (2011). *Metal and Metalloid Contaminants in Atmospheric Aerosols from Mining Operations.* 221(1-4), 145–157. doi:10.1007/s11270-011-0777-x.

Csavina, J., Taylor, M. P., Félix, O., Rine, K. P., Eduardo Sáez, A., & Betterton, E. A. (2014). Size-resolved dust and aerosol contaminants associated with copper and lead smelting emissions: implications for emission management and human health. *The*

Science of the total environment, 493, 750–756. https://doi.org/10.1016/j.scitotenv.2014.06.031.

Dobson, C. M., Ellison, G. B., Tuck, A. F., & Vaida, V. (2000). Atmospheric aerosols as prebiotic chemical reactors. *Proceedings of the National Academy of Sciences of the United States of America*, *97*(22), 11864–11868. https://doi.org/10.1073/pnas.200366897.

Duarte, R. M. B. O. and Duarte, A. C. (2021). On the Water-Soluble Organic Matter in Inhalable Air Particles: Why Should Outdoor Experience Motivate Indoor Studies? *Appl. Sci.,* 11, 9917. https://doi.org/10.3390/ app11219917.

Ellison, G. B., Tuck, A. F., Vaida, V. (1999). Atmospheric processing of organic aerosols. *Journal of Geophysical Research,* 104(D9), 11633–. doi:10.1029/1999jd900073.

Farah, A. (2018). Analysis of the physical and chemical properties of atmospheric aerosol at the Puy de Dôme station. *Earth Sciences.* Université Clermont Auvergne.

Hays, M. (2014). Physical and Chemical Properties of Anthropogenic Aerosols: An Overview. *Presented at Goldschmidt Conference,* Sacramento, CA.

Haque, M. M., Kawamura, K., Dhananjay, K. D., Cao, F., Wenhuai, S., Bao, M., and Zhang Y. (2019). Characterization of organic aerosols from a Chinese megacity during winter: predominance of fossil fuel combustion. *Atmos. Chem. Phys.,* 19, 5147–5164. https://doi.org/10.5194/acp-19-5147-2019.

Heald, C. L., and Kroll, J. H. (2020). The fuel of atmospheric chemistry: Toward a complete description of reactive organic carbon. *Science advances*, 6(6), eaay8967. https://doi.org/10.1126/sciadv.aay8967.

Hidy, G. M. (2019). Atmospheric Aerosols: Some Highlights and Highlighters, 1950 to 2018. *Aerosol Science and Engineering,* 3, 1-20. doi:10.1007/s41810-019-00039-0.

Huanga, Y., Mahrta, F., Xua S., Shiraiwac M., Zuendd, A. and Allan K. B. (2021). Coexistence of three liquid phases in individual atmospheric aerosol particles: *Proceedings of the National Academy of Sciences,* Edited by Thomas Peter, Institute for Atmospheric and Climate Science, 118(16), e2102512118. https://doi.org/10.1073/pnas.2102512118.

Jacobson, M. C.; Hansson, H.-C.; Noone, K. J.; Charlson, R. J. (2000). *Organic atmospheric aerosols: Review and state of the science. Reviews of Geophysics,* 38(2), 267–294. doi:10.1029/1998rg000045.

Jelena M. J., Marijana K. I., Snežana M., Aleksandra T., Tamara A., Jelena B. and Jasmina A. (2021) Assessing method performance for polycyclic aromatic hydrocarbons analysis in sediment using GC-MS: method validation and principal component analysis for quality control. *International Journal of Environmental Analytical Chemistry* 30(3), 1-16. doi:10.1080/23311843.2017.1339841.

Johnston, M. V. and Kerecman, D. (2019). Molecular Characterization of Atmospheric Organic Aerosol by Mass Spectrometry. *Annual Review of Analytical Chemistry,* 12(1), *annurev-anchem*-061516-045135. doi:10.1146/annurev-anchem-061516-045135.

Kanellopoulos, P. G., Verouti, E., Chrysochou, E., Koukoulakis, K. and Bakeas, E. (2021). Primary and secondary organic aerosol in an urban/industrial site: Sources, health implications and the role of plastic enriched waste burning. *Journal of Environmental Sciences, 99(), 222–238.* doi:10.1016/j.jes.2020.06.012.

Kalliat T. V., Franz S. E., Aubrey A. H. and Mickael V. (2015). Mass Transport and Chemistry at the Air–Water Interface of Atmospheric Dispersoids, Editor(s): Satinder Ahuja, *Food, Energy, and Water,* Elsevier, 93-112.

Kudom D. S. (2019). *Persistent Organic Pollutants* || Degradation Pathways of Persistent Organic Pollutants (POPs) in the Environment, 10.5772/intechopen.74146(Chapter 3), –. doi:10.5772/intechopen.79645.

Lepri, L., Del Bubba, M., Masi, F., Udisti, R. & Cini, R. (2000). Particle Size Distribution of Organic Compounds in Aqueous Aerosols Collected from Above Sewage Aeration Tanks, *Aerosol Science & Technology,* 32:5, 404-420, doi:10.1080/027868200303542.

Li, Z., Zhou, R., Li, Y., Chen, M., Wang, Y., Huang, T., Yi, Y., Hou, Z., Meng, J., Yan, L. (2021). Characteristics and Sources of Organic Aerosol Markers in $PM_{2.5}$. *Aerosol Air Qual. Res.* 21, 210180. https://doi.org/10.4209/aaqr.210180.

Manisalidis, I., Stavropoulou, E., Stavropoulos, A. and Bezirtzoglou, E. (2019). Environmental and Health Impacts of Air Pollution: A review. *Front. Public Health* 8:14. doi:10.3389/fpubh.2020.00014.

McFiggans, G., Mentel, T. F., Wildt, J., Pullinen, I., Kang, S., Kleist, E., Schmitt, S., Springer, M., Tillmann, R., Wu, C., Zhao, D., Hallquist, M., Faxon, C., Le Breton, M., Hallquist, A. M., Simpson, D., Bergström, R., Jenkin, M. E., Ehn, M., … Kiendler-Scharr, A. (2019). Secondary organic aerosol reduced by mixture of atmospheric vapours. *Nature.* 565(7741):587-593. doi:10.1038/s41586-018-0871-y.

McKay, C. P., Coustenis, A., Samuelson, R. E., Lemmon, M. T., Lorenz, R. D., Cabane, M., Rannou, P. and Drossart, P. (2001). *Physical properties of the organic aerosols and clouds on Titan.,* 49(1), 79–99. doi:10.1016/s0032-0633(00)00051-9.

O'Brien, R. E., Neu, A., Epstein, S. A., MacMillan, A. C., Wang, B., Kelly, S. T., Nizkorodov, S. A., Laskin, A., Moffet, R. C., and Gilles, M. K. (2014), Physical properties of ambient and laboratory-generated secondary organic aerosol, *Geophys. Res. Lett.,* 41, 4347– 4353, doi:10.1002/2014GL060219.

Patel, A. B., Shaikh, S., Jain, Kunal R., Desai, C. and Madamwar, D. (2020). Polycyclic Aromatic Hydrocarbons: Sources, Toxicity, and Remediation Approaches. *Frontiers in Microbiology, 11(), 562813–.* doi:10.3389/fmicb.2020.562813

Penkała, M., Ogrodnik, P. and Rogula-Kozłowska, W. (2018). Particulate Matter from the Road Surface Abrasion as a Problem of Non-Exhaust Emission Control. *Environments*, *5*, 9. https://doi.org/10.3390/environments5010009.

Rinaldi, M., Decesari, S., Finessi, E., Giulianelli, L., Carbone, C., Fuzzi, S., O'Dowd, C. D., Ceburnis, D., & Facchini, M. C. (2010). Primary and Secondary Organic Marine Aerosol and Oceanic Biological Activity: Recent Results and New Perspectives for Future Studies. *Advances in Meteorology,* 20(12), 1–10. doi:10.1155/2010/310682.

Rushdi, A. I., El-Mubarak, A. H., Lijotra, L., Al-Otaibi, M. T., Qurban, M. A., Al-Mutlaq, K. F. and Simoneit, B. R. T. (2014). Characteristics of organic compounds in aerosol particulate matter from Dhahran city, Saudi Arabia. *Arabian Journal of Chemistry, 55, S1878535214000550–.* doi:10.1016/j.arabjc.2014.03.001.

So H. J., Hyung B. L., Na R. C., Ji Y. L., Yun K. A. and Yong P. K. (2019). Classification and Characterization of Organic Aerosols in the Atmosphere over Seoul Using Two Dimensional Gas Chromatography-time of Flight Mass Spectrometry (GC×GC/TOF-

MS) Data. *Asian Journal of Atmospheric Environment,* 13,(2), 88-98. doi: https://doi.org/10.5572/ajae.2019.13.2.088.

Song, C., Na, K. and Cocker, D. R. (2005). Impact of the Hydrocarbon to NOx Ratio on Secondary Organic Aerosol Formation. *Environmental Science & Technology, 39(9), 3143–3149.* doi:10.1021/es0493244.

Su, H., Cheng, Y., & Pöschl, U. (2020). New Multiphase Chemical Processes Influencing Atmospheric Aerosols, Air Quality, and Climate in the Anthropocene. *Accounts of chemical research, 53*(10), 2034–2043. https://doi.org/10.1021/acs.accounts.0c00246.

Tefera, W., Kumie, A., Berhane, K., Gilliland, F., Lai, A., Sricharoenvech, P., Patz, J. Samet, J. and Schauer, J. J. (2021). Source Apportionment of Fine Organic Particulate Matter (PM2.5) in Central Addis Ababa, Ethiopia. *Int. J. Environ. Res. Public Health* 2021, 18, 11608. https:// doi.org/10.3390/ijerph182111608.

Tegen, I. and Schepanski, K. (2018). Climate Feedback on Aerosol Emission and Atmospheric Concentrations. *Curr Clim Change Rep* 4, 1–10. https://doi.org/10.1007/s40641-018-0086-1.

Thakur, M. and Deepak P. (2020). Chapter 12 - Environmental fate of organic pollutants and effect on human health, Editor(s): Pardeep Singh, Ajay Kumar, Anwesha Borthakur, *Abatement of Environmental Pollutants,* Elsevier, 245-262, https://doi.org/10.1016/B978-0-12-818095-2.00012-6.

Viana, M.; Pey, J.; Querol, X.; Alastuey, A.; de Leeuw, F.; Lükewille, Anke (2014). Natural sources of atmospheric aerosols influencing air quality across Europe. *Science of The Total Environment, 472, 825–833.* doi:10.1016/j.scitotenv.2013.11.140.

Yiyi W., Tingting C. and Jonathan E. T. (2012). The chemical evolution & physical properties of organic aerosol: A molecular structure based approach. *Atmospheric Environment,* 62(none), –. doi:10.1016/j.atmosenv.2012.08.029.

Zhu, L., Henze, D. K., Bash, J. O., Cady-Pereira, K. E., Shephard, M. W. Luo, M. and Capps, S. L. (2015). Sources and Impacts of Atmospheric NH3: Current Understanding and Frontiers for Modeling, Measurements, and Remote Sensing in North America. *Current Pollution Reports,* 1(2), 95–116. doi:10.1007/s40726-015-0010-4.

Chapter 4

Atmospheric Bioaerosol

Shraddha Sanjiv Shirsat[1],
Sharad Chandrakant Gangavane[1,*],
Vijay Jagdish Upadhye[2,*],
Babasaheb Shivmurti Surwase[2]
and Gauri Yogendra Kulkarni[3]

[1]Rajiv Gandhi Institute of IT and Biotechnology (RGITBT),
Bharati Vidyapeeth Deemed University, Katraj, Pune, Maharashtra, India
[2]Junior Research Fellow Biotechnology, School of Life Sciences, SRTM University,
Nanded, Maharashtra, India
[3]Department of Microbiology, Parul Institute of Applied Sciences (PIAS),
Parul University, Vadodara, Gujarat, India
[4]Department of Botany, School of Life Sciences, SRTM University,
Nanded, Maharashtra, India
[5]Department of Biotechnology, (DSCL), SRTM-University,Nanded,
Maharashtra, India

Abstract

Air is a mixture of gases and other constituents, including biological organisms and cell debris. Recent pandemics and the rise of airborne ailments have highlighted bioaerosols vital role in combatting various infectious diseases, air pollution, and bioterrorism. The study of bioaerosols is also crucial for understanding the impacts of biogeochemical cycles, biodiversity, changes in ecosystems, air quality, and exposure in hospitals, forensic investigations, and industrial hygiene.

* Corresponding Author's Email: sharadbiotech1986@gmail.com.
* Corresponding Author's Email: dr.vijaysemilo@gmail.com.

In: Atmospheric Aerosols
Editor: Binoy K Saikia
ISBN: 979-8-88697-211-5

Bioaerosols known as biological aerosols or organic dust are tiny particles in the range of 0.001 μm to 100 μm floating in the atmosphere with biological origins. They can be categorized as viable and non-viable. The viable bioaerosols include fungi, algae, viruses, and bacteria and non-viable bioaerosols include proteins, endotoxins, and dead cells. There has been tremendous development in the tools for collecting air samples and analyzing bioaerosols that can prove to benefit economic and ecological aspects. Despite this, there is a lack of standardization in regards to the maximum limit of concentration of atmospheric bioaerosols and the absence of uniformity in approaches for collection, analysis, and detection.This chapter deals to provide an overview of sources, types, transmission and transport mechanisms, collection, measurement, protection, and control methods, the significance of atmospheric bioaerosols, and future research needed to assess bioaerosols effectively.

Keywords: bioaerosols, air-pollution, bioterrorism, biodiversity, economic andecological aspects, protection and control methods

Introduction

Air is considered one of the important sources of creation. Other sources of creation to be noted are water, soil, etc. Scientists had been trying to understand the components and properties of air. Air is an invisible layer of blanket around the surface of the earth but it holds hidden entities that are significant from a medical as well as environmental point of view. Air is a mixture of gases and other constituents, including biological organisms and cell debris. A suspension of fine solid particles or liquid droplets in air or another gas is known as an aerosol. Atmospheric aerosol is cited as particulate matter (PM) ranging from sizes of 2 nm –10 μm present within the ambient atmosphere. Aerosol particles can arise from human activities and natural processes or occasionally a mix. They'll be released directly or produced in situ because of chemical and physical alterations of gas-phase emissions within the atmosphere. Samples of natural aerosols are fog, mist, soil dust, volcanic dust, forest exudates, mineral dust, organic matter from biogenic carbon, and geyser steam. Samples of human-caused aerosols are soot, industrial aerosol, particulate air contaminants, and fume. Atmospheric aerosols could be present as liquid, solid, gas form, or liquid trapped on the solid particles. Individual atmospheric aerosol particles have a small size and hence little inertia, and thus they'll stay floating within the air for diverse days after being emitted naturally or formed by artificial activities within the atmosphere. As a result, during their atmospheric lifetime, they'll be transported long distances, participating in atmospheric chemical reactions and impacting human health, environmental quality, and climate [1, 2]. Aerosol generally refers to an aerosol spray that dispenses a consumer product from a container. The assembly and elimination of aerosols, industrial relevance of aerosols, consequences of aerosols on the surroundings, and others are covered in aerosol science. The spreading of pesticides, medical management of respirational ailments, and combustion equipment and tools are some technological applications of aerosols [3, 4, 5].

Bioaerosols known as biological aerosols are a subgroup of particles liberated from land and aquatic ecosystems into the environment. They contain both living and non-living components, like fungi, pollen, bacteria, endotoxins, exotoxins allergens, and viruses. The soil, water, wind, and living organisms like animals and plants are natural sources of bioaerosols. The anthropogenic sources of bioaerosol include sneezing and coughing, agriculture, wastewater treatment, and compost facilities. The bioaerosol generation or aerosolization into the atmosphere may be via wind turbulence over a surface. As soon as they are released within the surroundings, they'll be transmitted in the vicinity or worldwide: conventional wind patterns are accountable for local diffusion, while hot and humid storms and dust trails can move bioaerosols between landmasses. Above ocean layers, bioaerosols are produced using a sprayer and effervesces. Bioaerosols can transport various living entities like microbial pathogens, pollen, toxins, and allergens additionally with non-living components like dust, organic matter, soot, and other material to which humans are susceptible. An example of transmission and effects of bioaerosols on human health is the meningococcal meningitis outbreak in geographic regions linked to sand and dust waves during arid periods. Such other occurrences are associated with dust storms transmitting Mycoplasma and tuberculosis [6, 7].

The transport of dust particles into the atmosphere was initially detected by Darwin but Pasteur was the foremost to investigate microbes and their activeness in the air. Previously, laboratory cultures were accustomed to growing and isolating different bioaerosols as all the microbes cannot be cultured, many were undetected before the event of DNA-based tools. A process developed by Pasteur for sampling bioaerosols showed that more microbial activity occurred at lower altitudes and decreased at higher altitudes [8].

Today, the increase of pandemics and airborne ailments has highlighted the vital role of bioaerosols, this would increase the importance of studying bioaerosolswhich can help to combat various infectious diseases, pollution, and bioterrorism. The study of bioaerosols is additionally crucial for understanding the impacts of biogeochemical cycles, biodiversity, changes in ecosystems, air quality, and exposure in hospitals, forensic investigations, and industrial hygiene. There has been tremendous development in the tools for collecting air samples and analyzing bioaerosols that may benefit economic and ecological aspects. Despite this, there's a scarcity of standardization concerning the maximum limit of concentration of atmospheric bioaerosols and therefore the absence of uniformity in approaches for collection, analysis, and detection. This chapter deals to produce a summary of sources, types, transmission and transport mechanisms, collection, measurement, protection and control methods, the importance of atmospheric bioaerosols, and future research needed to assess bioaerosols effectively.

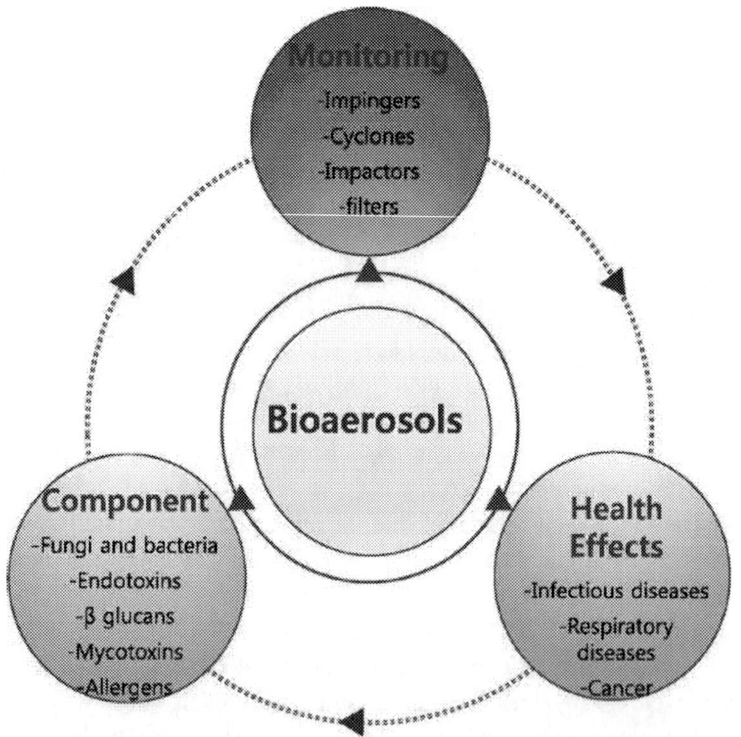

Figure 1. Schematic representation of Bioaerosols- Component, monitoring, and health effects [9].

Types of Bioaerosols

The components of bioaerosols comprise viruses, bacteria, protozoa, algae, and pollen. It has the highest concentrations in the planetary boundary layer (PBL) and decreases with altitude. Several biotic and abiotic factors influence the survival rate of bioaerosols which include climatic conditions, ultraviolet (UV) light, temperature, and humidity as well as resources present within dust or clouds [11].

Bacteria are the predominantly found bioaerosols component in the marine environments, whereas those found in the terrestrial atmosphere are rich in bacteria, fungi, and pollen. The dominance of particular bacteria and their nutrient sources are subject to change according to time and location [12]. According to time and location, the domination of particular bacteria and their nutrient resources changes [13].

The size of bioaerosols ranges from 10-nanometer virus particles to 100 micrometers of pollen grains [14]. Pollen grains are the largest bioaerosols and are less likely to remain suspended in the air over a long period due to their weight [15].

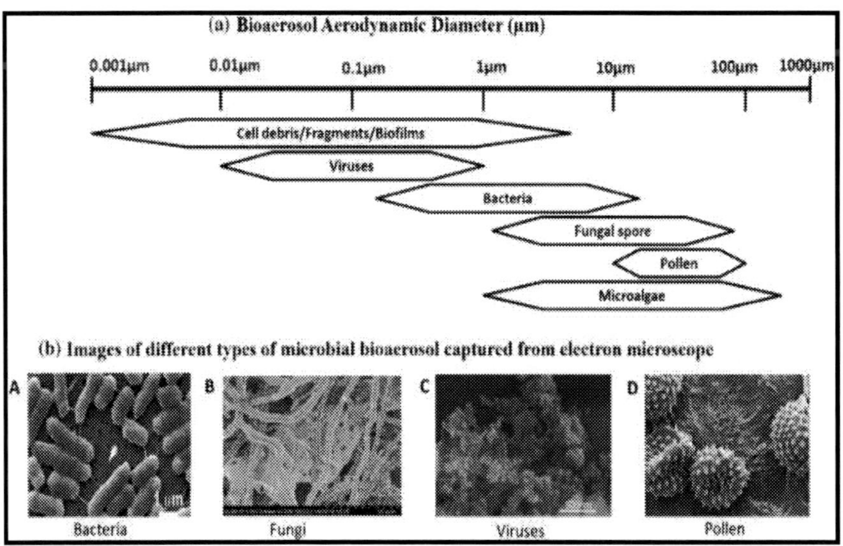

Figure 2. Types of bioaerosol components present in the atmosphere. (a) Bioaerosol aerodynamic diameter. (b) Images of different types of bioaerosol microorganisms view from an electron microscope [10].

Subsequently, pollen particle concentration decreases more rapidly with height than smaller bioaerosols such as bacteria, fungi, and possibly viruses, which may be able to survive in the upper troposphere. At present, there is little research on the specific altitude tolerance of different bioaerosols. However, scientists believe that atmospheric turbulence impacts where different bioaerosols may be found [12]. The aeration rate on the surface of the water, bubbles bursting, wind diffusion, and wind speed impact bioaerosol concentration containing microorganisms and other components due to changes in humidity, temperature, and geographic locations.

Fungi

Fungal cells usually die once they travel through the ecosystem due to the desiccating consequences of higher altitudes. But some particularly resilient fungal bioaerosols have been proven to continue to exist in atmospheric distribution positively under extreme UV to mild conditions [16]. Even though bioaerosol levels of fungal spores boom in better humidity situations, they also can be active in low humidity conditions and maximum temperature ranges fungal bioaerosols boom at extraordinarily low levels of humidity. The spores may be released by way of passive procedures, through the use of wind or other outside forces, or by using active techniques wherein an ejection is produced by way of, for example, osmotic stress or surface unease. It's far generally assumed that spores alongside fungal fragments constitute the bulk of the bioaerosols made out of fungi that are discovered within the size range of 1–50 μm, however, are found greater in size generally between 2 and 10 μm [21, 22]. Common genera within the atmosphere fungal spores are *Cladosporium*, *Alternaria,* and the actively launched ascospores and basidiospores[23].

Bacteria

Unlike other bioaerosols, microorganisms are capable to finish full reproductive cycles in the days or perhaps weeks that they live to tell the tale of the environment, making them a main issue of the air biota. These reproductive cycles aid a currently unproven idea that microorganism bioaerosols form communities in an atmospheric environment [7]. The survival of bacteria depends on water droplets from fog and clouds that offer bacteria vitamins and protection from UV light [12]. The acknowledged bacterial groupings which can be plentiful in antimicrobial environments around the arena include Bacillaceae, Actinobacteria, Proteobacteria, and Bacteroidetes [17].

Viruses

The air transports viruses and different pathogens considering that viruses are smaller than other bioaerosols, they can tour additional distances. In one simulation, a fungal spore and a virus have been concurrently launched from the pinnacle of a building; the spore traveled most effectively a hundred and fifty meters even as the virus traveled almost 200,000 horizontal kilometers [12].

In another study, aerosols (<5 μm) containing SARS-CoV-1 and SARS-CoV-2 were generated via an atomizer and fed into a Goldberg drum to create an aerosolized environment. The inoculum produced cycle thresholds between 20 and 22, much like the ones found in human upper and lower respiratory tract samples. SARS-CoV-2 remained potential in aerosols for 3 hours, with a lower contamination titer similar to SARS-CoV-1. The half-lives of both viruses in aerosols became 1.1 to at least 1.2 hours on average. The outcomes advise that the transmission of both viruses via aerosols is manageable, as they could remain possible and infectious in suspended aerosols for hours and on surfaces for up to days [18].

The literature on airborne viruses is relatively scarce which is maximum likely due to inefficient size techniques especially earlier than the development of molecular techniques inclusive of PCR. Viable viruses generally are transported to greater distances attached to larger debris as per the assumptions. Because viruses do not have any restore systems, they're effortlessly inactivated via warmth, radiation, and different types of environmental pressure than living microorganisms. The relationship between relative humidity and survival of viruses inside the air is complex and ranges relying on virus type [19].

Pollen

Despite being larger and heavier than other bioaerosols, some research shows that pollen may be transported over heaps of kilometers [12]. They're the main source of wind-dispersed allergens, coming particularly from seasonal releases from grasses and trees. The most common pollen allergens that are likely to cause allergic reactions are ragweed pollen, birch pollen, oak pollen, and pollen released by other anemophilous plants [6]. It has been observed that monitoring distance, transport, assets, and deposition of pollen to terrestrial and marine environments are beneficial for decoding pollen statistics [6]. The dispersal, resuspension, and shipping of pollen into the environment are largely decided by meteorological factors, for instance, emissions are lower when wind speeds are low and during rainfall. There also appears to be an association between transport distances and temperature, with a boom in distances of pollen traveled because the temperature increases [19].

β-glucans

There may be little or no facts available on the epidemic and health hazards related to glucans. However, the facts indicate that the β-glucans at once assault the lungs and suppress the immune process, and create a risk of irritation and sensitization. Similarly, contact with β-glucan is associated with respiratory symptoms, airway responsiveness, toxic pneumonitis, chronic bronchitis, and lung function decline. At the spin facet, it has business cost as a nutritional supplement that enhances the immune device and improves the healing recovery impact because of positively targeting of the immune cells [31].

Endotoxins and Mycotoxins

Endotoxins are big biomolecules made of lipids and polysaccharide compounds that might be determined in the outer membrane of the cell wall of gram-negative microorganisms and feature strong seasoned-inflammatory properties. They include numerous compounds like lipid elements, O-connected polysaccharide side chains (O-antigen), and middle polysaccharide chain (lipid A) that are liable for poisonous results. The complexity of endotoxin stems from the reality that it is not a single uniform substance that causes size uncertainties. The contact with the endotoxins are steady due to the fact they could bind with dirt particle and be effortlessly inhaled, as a result, they appear as one of the crucial compounds which contribute to organic dust resulting in toxic syndrome and occupational lung sickness. Additionally, publicity to endotoxin compounds purpose a decrease in lung characteristic and also cause health risks like fever, shivering, blood leukocytosis, arthralgia, dyspnea, etc.

Mycotoxins are fungal pollutants with a low molecular weight that originate from single mold species in addition to more than one mildew species and pose a higher level of poisonous risk to both human and animal species. As a result, mycotoxins are categorized based totally on their chemical structure and related reactive groups inclusive of amines, carboxylic acids, phenolics, lactams, and amides. Mycotoxins are risky to health and can cause immune system weakness, hypersensitive reactions or inflammation, or even demise in extreme instances. But, time of contact, the concentration of mycotoxins, and man or woman's conditions inclusive of age, gender and fitness status have a robust affect on the mycotoxicosis effect. Ingestion, spore inhalation, and dermal touch with mold-infested substrates are the maximum common routes of entry of mycotoxins [31]. The major mycotoxins that affect human health are aflatoxins, citrinin, ergot alkaloids, fumonisins, ochratoxins, etc.

Other Bioaerosols

Few studies have investigated airborne microalgae and cyanobacteria, but they have nevertheless been found in both indoor and outdoor air around the globe [23, 24]. A variety of biologically derived particles in the air can be added to the list. Crusts on rocks and sand often carry biological material that may be suspended in the air as a result of erosion from wind and rainfall [25]. Lichens, which are a symbiotic union between fungi and microalgae or cyanobacteria, are widespread on Earth and have also been detected in the atmosphere [26]. Plant fragments are, by mass, among the most common bioaerosol materials in the atmosphere [27]. Other particle types that can be important, especially from a health point of view as they may cause allergic reactions, are skin fragments, fur fibers, insect secretions, and dandruff [28].

There are numerous sorts of microorganisms in air however air isn't always the natural habitat of all forms of microorganisms because it does not constitute the best moisture and nutrient for the increase of microorganisms. Hence, bioaerosols sources can be varied such as natural or anthropogenic.

Table 1. Sources of bioaerosols [27]

Bioaerosol source	Mechanisms
Healthcare	Surgical or dental procedures
	Hospital air
	Mechanical ventilators, bed making, and resuspension on dust or skin squamae
Water industry	Cooling towers
	Wastewater irrigation sites
Agricultural/forestry industries	Grain harvesting, food processing, dust, and/or feces from animal housing and farming activities
	Insecticidal crop spraying
	Genetic dispersion
Postal and shopping industry	Mail sorting and opening
	Mist machine
Leisure activities	Marine activities, e.g., surfing
	Whirlpools
Human activity	Breathing
	Speaking
	Shouting
	Coughing
	Sneezing
	Vuvuzela playing
	Showering

Pollen, spores, indigenous microorganisms such as bacteria, viruses, algae, fungus, and so on are the natural sources. The essential anthropogenic foundations of bioaerosols are waste disposal region (which includes, landfill, organic waste treatment facilities, wastewater remedies, and many others.), cattle, horticulture, refrigeration centers, etc. [29]. Other than this, the nasal discharge from infected sufferers is some other primary source of bioaerosols. The pathogens can be released into the surroundings during the sneezing and coughing of the sufferers with the discharged release [47]. The following table shows in detail different sources of bioaerosols, their mechanism of distribution, and particle size range.

Routes of Exposure

Airborne microorganisms can be transported and infect human beings directly through the respiratory pathway, through dermal contact, or by ingestion. Besides, they can indirectly enter the human body during the handling of contaminated waste materials. The below figure depicts different types of bioaerosols and their routes of entry along with the harmful effects.

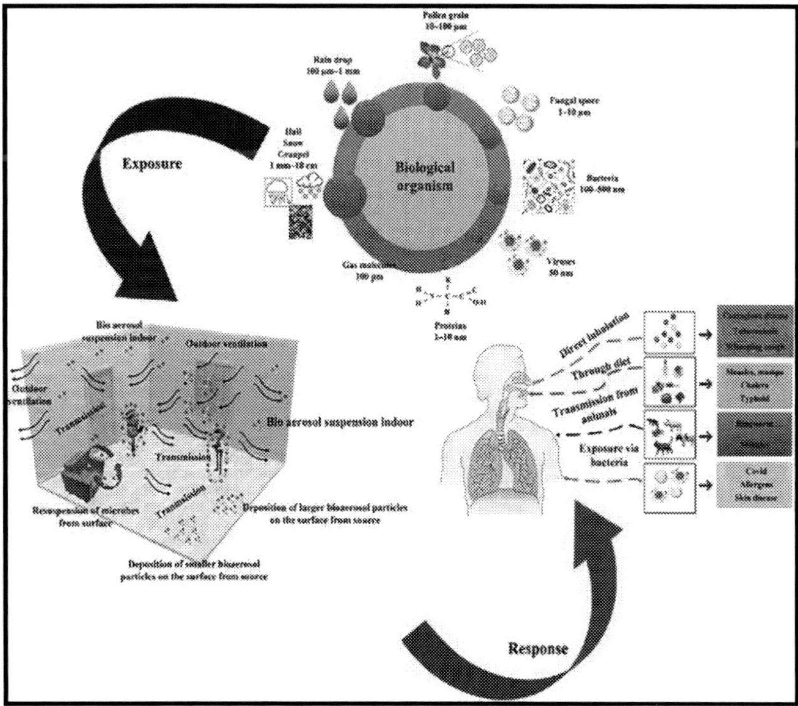

Figure 3. Route of entry of bioaerosols by various sources [29].

Elements Determining the Survival of Bioaerosols in Ambient Air

The survival of airborne microorganisms relies upon diverse environmental factors which include relative humidity (RH), solar radiation, and temperature. Regardless of the categories, environmental exposure is unfavorable to airborne microorganisms [32].

Humidity

The effect of relative humidity on airborne microorganisms turned into a study many years ago. It has been found that the inactivation of airborne microorganisms takes place in most quantities at around 50% relative humidity. The reasoning at the back of this can be acute dehydration of the cells and improvement of hypersensitive reaction of the cells to pollutants [33].

Solar Radiation

Ultraviolet radiation of daylight can probably damage airborne microorganisms [34]. The nucleic acids inside the dwelling cells can take in the radiation of wavelengths ranging from 200 nm to 300 nm. The power can damage the structure of the nucleic acids and pyrimidine dimers resulting in a change in shape that can avert the DNA replication technique and prevent protein formation which leads to the demise of the microorganism. Besides, the effects of UV radiation on microorganisms are related to the relative humidity of the ecosystem. UV inactivation of microorganisms decreases with the boom in relative humidity [35].

Temperature

The effect of ambient temperature on airborne microorganisms varies broadly with the species of the microorganisms. The thermophilic microorganisms can live to tell the tale in the temperature variety of 55-85°C. Therefore, it reflects that these organisms can't live in low-temperature regions. A totally low ambient temperature inhibits the boom of microbes [47].

Fate and Delivery of Bioaerosols in the Environment

Aero microorganisms are transported into the human body and environment by way of various dealers, pathogens, and other kinds. The three essential routes are Discharging, distribution, and deposition [36-40].

In launching, microbe debris gets suspended within the earth's atmosphere through wind turbulence over a surface. Inside the air, they generally stay within the planetary boundary layer (PBL), but in some cases reach the top layer known as the troposphere and stratosphere [47]. In this system, each terrestrial and aquatic asset of microorganisms is worried. Loading of microorganisms occurs to a higher degree from terrestrial sources than aquatic resources. They may be transported domestically or globally: routine wind styles and strengths are liable for local dispersal; at the same time it is observed that tropical storms and dust plumes can move bioaerosols among continents [7]. Over ocean surfaces, bioaerosols are generated via sea spray and bubbles. The airborne debris may get released and transported with the aid of wind present from numerous point sources, areas, or line sources.If the dispersion occurs via point sources then it may be labeled as instant point bases and continuous point bases. On the opposite, line resources and place assets are normally larger sites of pollutant launch. The transport of airborne microbes is dependent on the spatial and temporal variety [41]. The transportation over a short period (much less than 10 mins) and short distance (less than 100 m) is described as sub-micro-scale delivery. The micro-scale delivery is typically discovered inside small buildings or in a small enclosed space. The delivery over 10 minutes to 1 hour and a distance of 100 m to 1 km is called microscale transport. Mesoscale shipping is the transport of airborne microbes over long days) and long distance (extra than a hundred km or extra). A maximum of the airborne microbes is not able to survive for a long term inside the ecosystem, the most extensively used scales taken into consideration for aerobiological studies are submicron level and micro degree. But some viruses and bacteria had been observed to be transported on mesoscale and microscale [47].

A bioaerosol deposition system is a kind of process wherein the microbe-loaded particles adhere to the surface. Bioaerosols are used as a nucleus by cloud droplets, ice crystals, and precipitation wherein water or crystals can form or preserve on their floor. These interactions display that air debris can bring changes in the hydrological cycle, weather situations, and weathering around the world. The modifications can lead to consequences consisting of desertification that is magnified by weather shifts. Bioaerosols also intermix when pristine air and smog meet, changing visibility and air high-quality. At the same time as bioaerosols may additionally tour thousands of kilometers before deposition, their last distance of tour and path is dependent on meteorological, bodily, and chemical factors. The 3 essential processes for bioaerosol deposition are Gravity settling, floor impaction, and rain deposition [47].

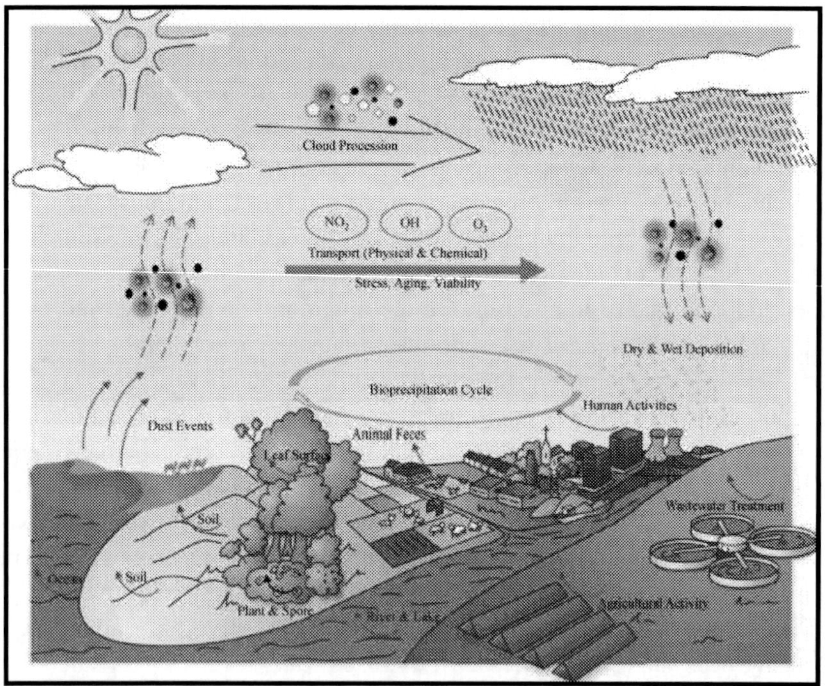

Figure 4. Sources and transport of bioaerosols [43].

All through the atmospheric shipping, bioaerosols may additionally interact with ultraviolet (UV) radiation, photooxidants, and various air pollution those interactions may result in additional chemical and bodily changes and organic changes leading to bioaerosols getting older in the course of delivery. When bioaerosols are uncovered to atmospheric oxidants (including OH, NO_3, and O_3) within the air, they rapidly trade from their authentic emission state. These oxidants can drastically alter the composition of bioaerosols by using homogeneous and heterogeneous chemical reactions. As an example, a direct assault by using reactive oxygen species could harm the organic systems but aggregation of cells or protecting envelopes may influence the viability of bioaerosols. Some noticeably pressure-resistant microorganisms observed within the deserts can thrive in extreme environmental conditions (loss of water, intense temperatures, and UV radiation). Intense weather conditions such as rain, snow, or dust storms and scavenging microorganisms from the atmosphere get deposit on the ground to influence the neighborhood microbial community structure [44].

Mechanisms to Control Bioaerosols

The control of aero microbe particles mechanisms used is to control the airflow of the units, filtration of the air, UV remedy, use of chemical products for attenuating the microorganisms, and isolating the contaminated space physically [38].

Ventilation

Proper air flow of the tainted areas is a common method used to manipulate the spread of airborne pathogens. In ventilation, air drift is created through regions wherein airborne microbe infection is maximum. In this method facilitating air movement inside the room is vital with the intention to enable dispersion of microorganisms away from the area of concern.

Filtration

Filtration is a powerful means to manipulate microbe contamination by installing HEPA filters or excessive-efficiency particulate air filters. HEPA filters are commonly installed on biosafety shelves. However, as their price is considered excessive, they are now not usually used as common air filters. The most normally used filtration system is the air filtration device that relies especially on the principles of baghouse filtration that works precisely similar to the vacuum cleaner bag.

Biocidal Dealers

These are used for the extra treatment of air and they may be employed for elimination of most of the airborne microbes with an affirmation of their nonviability and non-infectiousness.

Ultraviolet Germicidal Irradiation (UVGI)

The most typically employed, is the Ultraviolet Germicidal Irradiation method (UVGI). This biocidal agent can govern variegated airborne microbes however some of them have diverse stages of resistance in opposition to it.

Physical Isolation

Bodily isolating is defined because the barrier is created between distinctive compartments of surroundings through developing upbeat or downbeat pressure gradients of air and the use of airtight seals. Whilst the negative isolation method is used, the air is flown into the isolation chamber to guard the people out of the doors of the chamber. However, in the positive isolation method, the air is flown out of the chamber to keep people away from the infection within the isolation chamber.

Other Control Measures

Control measures of bioaerosols may additionally involve dehydration, UV irradiation, high-quality heating, and ozonation [47].

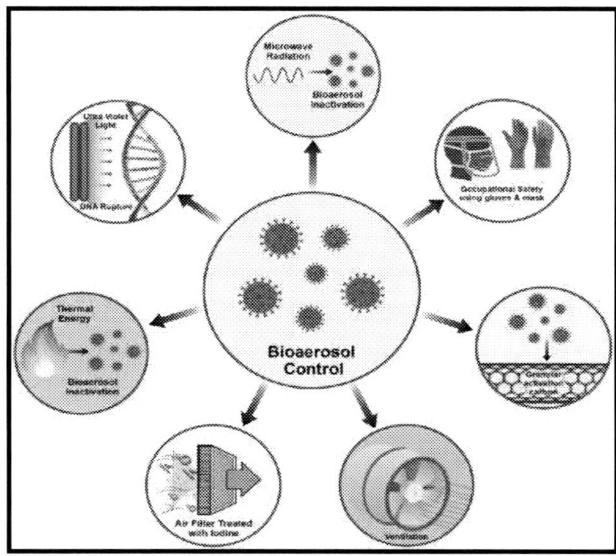

Figure 5. Bio-control measures of bioaerosols [45].

Collection and Identification of Bioaerosols

Bioaerosols are collected mainly by using collection plates, electrostatic collectors, mass spectrometers, and impactors. Polycarbonate filters have had the most correct bacterial sampling achievement.

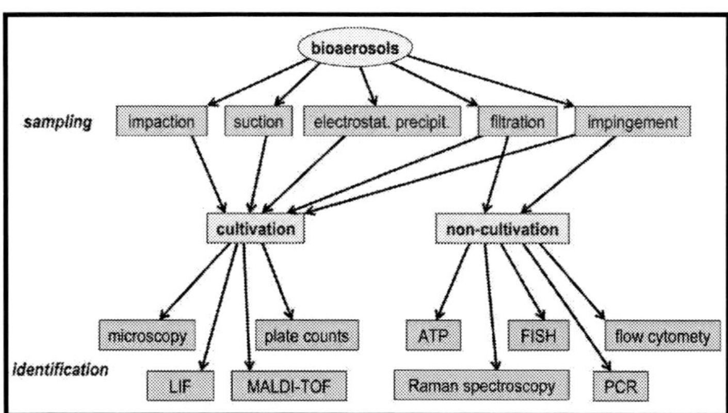

Figure 6. Collection and identification of bioaerosols [46].

Impactor

Impaction is used to split a particle from a stream primarily based on the inertia of the particle. An impactor consists of a series of nozzles (circular- or slot-shaped) and a goal. Best Impactor has a sharp cut-off or step-function efficiency curve. The debris larger than a selected aerodynamic size may be impacted onto a collection floor while smaller particles proceed through the sampler. Marple and Willeke have mentioned that high speed, inlet losses, interstage losses, and particle re-entrainment affect the overall performance traits of an impactor. The impactors can be of type's single-degree impactors and cascade impactors [54-55].

Impingers

Impingers are specially designed bubble tubes used for gathering airborne compounds right into a liquid medium. With impinger sampling, a recognized extent of air is bubbled through the impinger containing a distinct liquid. Rather than collecting onto a greased substrate or agar plate, impingers had evolved to impact bioaerosols into liquids, which includes deionized water or phosphate buffer. Commercially impingers encompass the AGI-30 (Ace Glass Inc.) and Bio sampler (SKC, Inc.).

Electrostatic Precipitators

Electrostatic precipitators, ESPs, have recently won renewed interest for bioaerosol sampling due to their enormously green particle removal efficiencies and gentler sampling technique as compared with impinging. ESPs charge and eliminate incoming aerosol debris from an airflow with the aid of employing a non-uniform electrostatic area among electrodes, and excessive field power. An electrostatic precipitator (ESP) removes debris from a gas circulation via the use of electricity to separate debris both definitely and negatively. The charged debris is then drawn to collector plates wearing the opposite charge.

In view that organic debris is commonly analyzed, the use of liquid-based assays (PCR, immunoassays, viability assay) is most well-known to pattern immediately right into a liquid volume for downstream evaluation.

Filters

Filters are often used to gather bioaerosols due to their simplicity and occasional cost. Filter collection is especially beneficial for personal bioaerosol sampling on account that they're mild and discreet. Filters have a size-selective inlet, including a cyclone or impactor, to dispose of larger debris and offer size grouping to the bioaerosol debris. Aerosol filters are often defined by the use of the period pore size or equivalent pore diameter. The word filter pore size does now not suggest the minimum particle size to be able to be gathered by way of the filter; in fact, aerosol filters commonly will collect debris a good deal smaller than the nominal pore size [50-52].

The Series of particles from a no biological aerosol sample is most normally accomplished by way of filtration. The filter media are to be had in each fibrous (commonly glass) and membranous system. Deposition occurs when debris impact and is intercepted through the fibers or floor of filters accordingly, debris smaller than the pore length may be successfully accrued. Sampling clear-out media may additionally have pore sizes of 0.01 to 10 μm [53, 56].

Cyclone

Cyclone captures bioaerosol right into a liquid matrix using centrifugal pressure. Samples may be analyzed with traditional structured and independent strategies. The most important risks are the viable evaporation of the liquid medium and the subsequent lack of sample and the low amount of aerosol debris quantity amassed. Furthermore, larger molecules are preferentially amassed. The advantage of this sampler is that it is easy to sterilize. However, in recent times, researchers are looking to improve real-time bioaerosol monitoring [49, 50, 57].

Conclusion

Bioaerosols are becoming a mode of biological attack making their way into war zones. This makes the study of bioaerosols of utmost importance as it can create bioterrorism and antisocial activities across the globe. There are no validated standard methods for measuring the number of bioaerosols in the atmosphere hence there is still scope for research in this arena. Moreover, there are no occupational exposure limits for the concentration of bioaerosols in the atmosphere. It is very important to regulate the exposure limits of bioaerosols as the exposure to viable bioaerosols such as bacteria, fungi, viruses, and toxins are known to cause various diseases such as infectious disease, acute respiratory disease, allergies, cancer, neurological problems, and birth defects. Methods to assess bioaerosol exposures are available; however, the selection of the most appropriate methods is highly dependent on the specific goals of the study. Therefore, more research is needed to establish better exposure assessment tools and to validate newly developed methods.

References

[1] Hinds, William C. *Aerosol technology: properties, behavior, and measurement of airborne particles.* John Wiley & Sons, 1999.

[2] Hidy, G. *Aerosols: An industrial and environmental science.* AcademicPress, New York.1984. e187.

[3] McNeill, V. Faye. Atmospheric aerosols: clouds, chemistry, and climate. *Annual review of chemical and biomolecular engineering* 8, 2017: 427-444.

[4] Colbeck, Ian, and MihalisLazaridis, eds. *Aerosol Science: technology and applications.* John Wiley & Sons. 2014.

[5] Fuller, Joanna Kotcher. E-Book:*SurgicalTechnology: Principles and Practic*e. Elsevier Health Sciences. 2017. ISBN: 9780323430562.

[6] Fröhlich-Nowoisky, J., Kampf, C. J., Weber, B., Huffman, J. A., Pöhlker, C., Andreae, M. O., Lang-Yona, N., Burrows, S. M., Gunthe, S. S., Elbert, W., Su, H., Hoor, P., Thines, E., Hoffmann, T., Després, V. R., &Pöschl, U. Bioaerosols in the Earth system: Climate, health, and ecosystem interactions. *AtmosphericResearch* 182 (2016): 346-376.

[7] Smets, Wenke, Serena Moretti, Siegfried Denys, and Sarah Lebeer. Airborne bacteria in the atmosphere: Presence, purpose, and potential. *Atmospheric Environment* 139, 2016: 214-221.

[8] Darwin, Charles. An account of the Fine Dust which often falls on Vessels in the Atlantic Ocean. *Quarterly Journal of the Geological Society.* 2, no. 1-2, 1846: 26-30.

[9] Kim, Ki-Hyun, EhsanulKabir, and ShaminAraJahan. Airborne bioaerosols and their impact on human health. *Journal of Environmental Sciences* 67, 2018: 23-35.

[10] Ghimire, Prakriti Sharma, LekhendraTripathee, Pengfei Chen, and Shichang Kang. Linking the conventional and emerging detection techniques for ambient bioaerosols: a review. *Reviews in Environmental Science and Bio/Technology* 18, no. 3, 2019: 495-523.

[11] Acosta-Martinez, Veronica, Scott Van Pelt, J. Moore-Kucera, M. C. Baddock, and Ted M. Zobeck. Microbiology of wind-eroded sediments: Current knowledge and future research directions. *Aeolian Research* 18, 2015: 99-113.

[12] Núñez, Andrés, Guillermo Amo de Paz, Alberto Rastrojo, Ana M. García, Antonio Alcamí, A. Montserrat Gutiérrez-Bustillo, and Diego A. Moreno. Monitoring of airborne biological particles in outdoor atmosphere. Part 1: Importance, variability and ratios. *International Microbiology* 19 no 1 2016: 1-13.

[13] Smets, Wenke, Serena Moretti, Siegfried Denys, and Sarah Lebeer. Airborne bacteria in the atmosphere: Presence, purpose, and potential. *Atmospheric Environment* 139 (2016): 214-221.

[14] Hitz, Carmen, and Walter Krebs. Short-term dynamic patterns of bioaerosol generation and displacement in an indoor environment. *Aerobiologia* 24 (2008): 203-209.

[15] Tang, Julian W. The effect of environmental parameters on the survival of airborne infectious agents. *Journal of the Royal Society Interface* 6, no. suppl_6 (2009): S737-S746.

[16] Dasgupta, Purnendu K., and Simon K. Poruthoor. Automated measurement of atmospheric particle composition. In *Comprehensive Analytical Chemistry*, vol. 37, pp. 161-218. Elsevier, 2002.

[17] van Doremalen, N., Bushmaker, T., Morris, D. H., Holbrook, M. G., Gamble, A., Williamson, B. N., Tamin, A., Harcourt, J. L., Thornburg, N. J., Gerber, S. I., Lloyd-Smith, J. O., de Wit, E., & Munster, V. J.Aerosol and surface stability of SARS-CoV-2 as compared with SARS-CoV-1. *New England journal of medicine* 382, no. 16 (2020): 1564-1567.

[18] Tang, Julian W. The effect of environmental parameters on the survival of airborne infectious agents. *Journal of the Royal Society Interface* 6, no. suppl_6 (2009): S737-S746.

[19] Kuparinen, Anna, Gabriel Katul, Ran Nathan, and Frank M. Schurr. Increases in air temperature can promote wind-driven dispersal and the spread of plants. *Proceedings of the Royal Society B: Biological Sciences* 276, no. 1670 (2009): 3081-3087.

[20] Elbert, W., P. E. Taylor, M. O. Andreae, and U. Pöschl. Contribution of fungi to primary biogenic aerosols in the atmosphere: wet and dry discharged spores, carbohydrates, and inorganic ions. *Atmospheric Chemistry and Physics* 7, no. 17 (2007): 4569-4588.

[21] Reponen, Tiina, Anne Hyvärinen, JuhaniRuuskanen, TaistoRaunemaa, and AinoNevalainen. Comparison of concentrations and size distributions of fungal spores in buildings with and without mold problems. *Journal of Aerosol Science* 25, no. 8 (1994): 1595-1603.

[22] Górny, Rafał L., TiinaReponen, Klaus Willeke, DetlefSchmechel, EnricRobine, Marjorie Boissier, and Sergey A. Grinshpun. Fungal fragments as indoor air biocontaminants. *Applied and environmental microbiology* 68, no. 7 (2002): 3522-3531.

[23] Genitsaris, Savvas, Konstantinos Ar Kormas, and Maria Moustaka-Gouni. Airborne algae and cyanobacteria: occurrence and related health effects. *Frontiers in Bioscience* 3, no. 2 (2011): 772-87.

[24] Sharma, Naveen Kumar, Ashwani Kumar Rai, Surendra Singh, and Richard Malcolm Brown Jr. Airborne algae: Their present status and relevance 1. *Journal of Phycology* 43, no. 4 (2007): 615-627.

[25] Tegen, Ina, and Inez Fung. Modeling of mineral dust in the atmosphere: Sources, transport, and optical thickness. *Journal of Geophysical Research: Atmospheres* 99, no. D11 (1994): 22897-22914.

[26] Marshall, William A. Aerial dispersal of lichen soredia in the maritime Antarctic. *New Phytologist* 134, no. 3 (1996): 523-530.

[27] Després, VivianeR., Huffman, J. A., Burrows, S. M., Hoose, C., Safatov, AleksandrS., Buryak, G., Fröhlich-Nowoisky, J., Elbert, W., Andreae, MeinratO., Pöschl, U., &Jaenicke, R.Primarybiologicalaerosolparticles in the atmosphere: a review. *Tellus B: Chemical and Physical Meteorology* 64, no. 1 (2012): 15598.

[28] Thomas, Richard James. Particle size and pathogenicity in the respiratory tract. *Virulence* 4, no. 8 (2013): 847-858.

[29] Kummer, Volker, and Wolf R. Thiel. Bioaerosols–sources and control measures. *International journal of hygiene and environmental health* 211, no. 3-4 (2008): 299-307.

[30] Gollakota, Anjani RK, SnehaGautam, M. Santosh, Harihara A. Sudan, Rajiv Gandhi, Vincent Sam Jebadurai, and Chi-Min Shu. Bioaerosols: Characterization, pathways, sampling strategies, and challenges to geo-environment and health. *Gondwana Research* (2021).

[31] Gollakota, Anjani RK, SnehaGautam, M. Santosh, Harihara A. Sudan, Rajiv Gandhi, Vincent Sam Jebadurai, and Chi-Min Shu. Bioaerosols: Characterization, pathways, sampling strategies, and challenges to geo-environment and health. *Gondwana Research* (2021).

[32] Tang, Julian W. The effect of environmental parameters on the survival of airborne infectious agents. *Journal of the Royal Society Interface* 6, no. suppl_6 (2009): S737-S746.

[33] Dunklin, Edward W., and Theodore T. Puck. The lethal effect of relative humidity on air-borne bacteria. *The Journal of experimental medicine* 87, no. 2 (1948): 87.

[34] Peccia, Jordan, Holly M. Werth, Shelly Miller, and Mark Hernandez. Effects of relative humidity on the ultraviolet induced inactivation of airborne bacteria. *Aerosol Science & Technology* 35, no. 3 (2001): 728-740.

[35] R.L. Riley, J. E. Kaufman, Effect of relative humidity on the inactivation of airborne Serratiamarcescens by ultraviolet radiation. *Applied and environmental microbiology* 23, 1972: 1113-1120.

[36] Dixit, Shraddha, and Prasenjit Adak. A Brief Note on Bioaerosols. *European Journal of Molecular & Clinical Medicine* 7, no. 7 (2020): 4301-4304.

[37] Mahesh, Koneti Venkata, Sachin Kumar Singh, and Monica Gulati. A comparative study of top-down and bottom-up approaches for the preparation of nanosuspensions of glipizide. *Powder technology* 256 (2014): 436-449.

[38] Vaid, Sachin Kumar, B. Kumar, A. Sharma, A. K. Shukla, and P. C. Srivastava. Effect of Zn solubilizing bacteria on growth promotion and Zn nutrition of rice. *Journal of soil science and plant nutrition* 14, no. 4 (2014): 889-910.

[39] Kumar, Vijay, Simranjeet Singh, Joginder Singh, and NirajUpadhyay. Potential of plant growth promoting traits by bacteria isolated from heavy metal contaminated soils. *Bulletin of environmental contamination and toxicology* 94, no. 6 (2015): 807-814.

[40] Mukherjee, R. Electrical, thermal and elastic properties of methylammonium lead bromide single crystal. *Bulletin of Materials Science* 43, no. 1 (2020): 1-5.

[41] National Research Council. *Biosolids applied to land: advancing standards and practices*. National Academies Press, 2002.

[42] Mahesh, KonetiVenkata, Sachin Kumar Singh, and Monica Gulati. A comparative study of top-down and bottom-up approaches for the preparation of nanosuspensions of glipizide. *Powder technology* 256 (2014): 436-449.

[43] Pepper, Ian L., and Charles P. Gerba. Aeromicrobiology. In *Environmental Microbiology*, pp. 89-110. Academic Press, 2015.

[44] Xie, Wenwen, Yanpeng Li, Wenyan Bai, JunliHou, Tianfeng Ma, Xuelin Zeng, Liyuan Zhang, and Taicheng An. The source and transport of bioaerosols in the air: A review. *Frontiers of Environmental Science & Engineering* 15, no. 3 (2021): 1-19.

[45] Smith, D. J., Thakrar, P. J., Bharrat, A. E., Dokos, A. G., Kinney, T. L., James, L. M., Lane, M. A., Khodadad, C. L., Maguire, F., Maloney, P. R., &Dawkins, N. L. A balloon-based payload for exposing microorganisms in the stratosphere (E-MIST). *Gravitational and Space Research* 2, no. 2 (2014).

[46] Mandal, Jyotshna, and Helmut Brandl. Bioaerosols in indoor environment-a review with special reference to residential and occupational locations. *The Open Environmental & Biological Monitoring Journal* 4, no. 1 (2011).

[47] Dixit, Shraddha, and Prasenjit Adak. A Brief Note on Bioaerosols. *European Journal of Molecular & Clinical Medicine* 7, no. 7 (2020): 4301-4304.

[48] Singh, Nitin Kumar, Gaurav Sanghvi, Manish Yadav, HirendrasinhPadhiyar, and ArtiThanki. Astate-of-the-art review on WWTP associated bioaerosols: Microbial diversity, potential emission stages, dispersion factors, and control strategies. *Journal of Hazardous Materials* (2020): 124686.

[49] Wang, Chi-Hsun, Bean T. Chen, Bor-Cheng Han, Andrew Chi-Yeu Liu, Po-Chen Hung, Chih-Yong Chen, and Hsing Jasmine Chao. Field evaluation of personal sampling methods for multiple bioaerosols. *PloS one* 10, no. 3 (2015): e0120308.

[50] Lindsley, William G., Brett J. Green, Francoise M. Blachere, Stephen B. Martin, Brandon F. Law, Paul A. Jensen, and Millie P. Schafer. Sampling and characterization of bioaerosols. *National Institute for Occupational Safety and Health* (2017).

[51] Mainelis, Gediminas, Klaus Willeke, AtinAdhikari, TiinaReponen, and Sergey A. Grinshpun. Design and collection efficiency of a new electrostatic precipitator for bioaerosol collection. *Aerosol Science & Technology* 36, no. 11 (2002): 1073-1085.

[52] Ladhani, Laila, Gaspard Pardon, Hanne Meeuws, Liesbeth van Wesenbeeck, Kristiane Schmidt, LievenStuyver, and Wouter van der Wijngaart. Sampling and

detection of airborne influenza virus towards point-of-care applications. *PloS one* 12, no. 3 (2017): e0174314.

[53] Lindsley, W. G. Filter pore size and aerosol sample collection. *NIOSH manual of analytical methods* 14 (2016).

[54] Hinds W. C., *Aerosoltechnology.* New York, NY: John Wiley & Sons, Ltd., 1982: 78, 104-126, 165.

[55] Marple, Virginia A., and Klaus Willeke. Inertial impactors: Theory, design, and use. *Fine Particles: Aerosol Generation, Measurement, Sampling, and Analysis* (1976): 412-446.

[56] Franchitti, Elena, Erica Pascale, ElisabettaFea, Elisa Anedda, and Deborah Traversi. Methods for bioaerosol characterization: limits and perspectives for human health risk assessment in organic waste treatment. *Atmosphere* 11, no. 5 (2020): 452.

[57] Douwes, J., Pl Thorne, N. Pearce, and D. Heederik. Bioaerosol health effects and exposure assessment: progress and prospects. *The Annals of occupational hygiene* 47, no. 3 (2003): 187-200.

Chapter 5

An Introduction to Bioaerosols: Properties and Behaviour

Nishi Srivastava*

Birla Institute of Technology, Mesra, Ranchi, Jharkhand, India

Abstract

The small solid or liquid particles suspended in the air are called aerosols, inhaled by a human while breathing. Even though being very small in size, aerosols play a vital role in human health and Earth's climate. Their sizes in the atmosphere vary from a few nanometers to micrometers, according to their origin and type. Aerosols present in the atmosphere have both natural and anthropogenic origins. Biological aerosols are usually called Bioaerosols. Bioaerosols are released into the atmosphere from the terrestrial or marine ecosystem. Bioaerosols are an essential component of the aerosol system in the atmosphere. They are less explored than their counterparts, such as dust aerosol, sulfate aerosol, black carbon aerosol, sea salt aerosol, etc. These bioaerosols consist of living and non-living organisms, including fungi, pollen, bacteria, viruses, and other biological fragments such as DNA fragments. Bioaerosols can transmit micro-organisms to humans, which can be allergic to human beings and cause severe illness. Bioaerosols enter the human body through inhalation and skin contact with air and various surfaces. Bioaerosols can cause an allergic response, toxic reactions, and infections in the human body. The bioaerosols became more relevant for study after a bio-terror attack in America in 2001 and influenza A H1N1 virus outbreak in 2009. The role of bioaerosols in transmitting the recent pandemic, *Coronavirus* disease 2019 (COVID-19), is also an important research area. World Health Organization (WHO) has announced the

* Corresponding Author's Email: nishi.bhu@gmail.com.

In: Atmospheric Aerosols
Editor: Binoy K Saikia
ISBN: 979-8-88697-211-5

possibility of the spread of COVID-19 through airborne transmission. Evidence indicates that aerosols contribute to the spread of COVID-19 under favorable conditions. The present chapter discusses the bioaerosol's origin, characteristics, behavior, distribution, and role in moderating human health and their possible role as agents in transmitting pandemics.

Keywords: bioaerosols, H1N1 virus, Coronavirus, COVID-19, pandemic

Introduction

Aerosols are suspended liquid and solid particles in the atmosphere with a wide range in size from sub-micron to several microns (Junge, 1963; Prospero et al., 1983). Natural and anthropogenic processes generate aerosols. Natural aerosols are injected into the atmosphere directly by volcanoes, burning of biomass, and fossil fuel, through the action of winds, biogenic aerosols, pollens, and also as a result of gas to particles conversion (Junge 1963; Prospero et al., 1983; Husar et al., 1997; Kaufman et al., 2002; 2005). Sea salt, dust, biomass burning aerosols, and volcanic aerosols are natural aerosols (Husar et al., 1997; Kaufman et al., 2002; 2005). Anthropogenic activities like fossil fuel burning, vehicular exhaust, industrial emission, mining, construction activities, transportation, modification in land cover, and so on generate anthropogenic aerosols in the atmosphere. Black carbon, volatile organic carbon, oxides of nitrogen and sulfur, and ozone are some examples of anthropogenic aerosols. The residence time of aerosols primarily depends on the size of aerosols. Aerosols larger in size (>100 μm) cannot remain suspended in the atmosphere for a long time and settle down under the effect of gravity. Aerosols smaller in size (sub-micron size) remain suspended in the atmosphere for a longer time (7-10 days) and can travel far from their source regions (Arimoto et al., 2001; Prospero et al., 2002; Gong et al., 2003; Maring et al., 2003). The distribution of aerosols is not uniform in the atmosphere and varies significantly both spatially and temporally. Regional distribution of aerosols (especially anthropogenic aerosols) is an essential input for the climate models and involves significant uncertainties (IPCC, 2013). Even though aerosols are primarily present in the atmospheric boundary layer, elevated aerosol layers are often found. The number concentration of aerosols depends on several factors such as location, atmospheric conditions, annual and diurnal cycles of meteorological parameters, and local sources of emissions.

On a global scale, natural aerosols constitute a substantial portion, but on a regional scale, anthropogenic aerosols may dominate natural aerosols (Satheesh and Moorthy, 2005). In urban areas, due to anthropogenic activities, the concentration of aerosols is usually much higher than in rural areas reaching 10^8 to 10^9 particles per cubic centimeter (Seinfeld and Pandis, 1998). Anthropogenic aerosols get transported long distances (due to their small size) and modify radiative properties at their source region and over areas where it is transported. During transit from one part to another, aerosols properties get modified. Unlike greenhouse gases, aerosols are highly heterogeneous in space and time. A high concentration of aerosols reduces the visibility of the atmosphere and causes other adverse effects on humans by affecting the respiratory system.

Aerosols are generated by several different processes occurring at the land surface, water body, and atmosphere. Aerosols are present in the lower layer of the atmosphere (i.e., troposphere) and the upper layer of the atmosphere (i.e., stratosphere). Still, there are notable differences in their sources, sizes, and chemical compositions. Aerosols occur in different shapes and are broadly classified as isometric particles, platelets, and fibers (Reist, 1984). Isometric particles are those for which all three dimensions are roughly the same, and one typical example is spherical particles. The second classification is platelets, particles with two long dimensions and one small third dimension; leaf fragments are examples of this type of aerosols. The third classification types, fibers, are the particles with great length in one dimension and much smaller dimensions in the other two. Threads or mineral fibers are examples of this type of aerosol. It should be noted that aerosols also occur in many different complex shapes along with these broad classifications (McCartney, 1976; Mishchenko et al., 1997).

Aerosols are classified into several groups depending on their generation, composition, and size. Based on the size of aerosols, three classifications of aerosols have been proposed in the literature (Junge, 1963). These classes are (i) Aitken mode or nucleation mode (0.001- 0.1 μm), (ii) Accumulation mode (0.1-1 μm), (iii) Coarse mode (>1 μm). The Aitken mode particles are the smallest aerosol with a diameter ranging between 0.001-0.1 μm. They are primarily formed by gas to particle conversion mechanisms. The particles in the atmosphere are produced by the nucleation of low volatile gases mainly produced because of industrial activities, forest fires, etc. This process is called a gas-to-particle conversion. In the accumulation mode, the particles grow by the coagulation of particles in Aitken modes and the condensation of vapor on the pre-existing particles. They are also emitted into the atmosphere by incomplete combustion of fossil fuels. Owing to its small size, the removal of the aerosols in this range is slow and tends to accumulate in the atmosphere until they are removed through wet removal. The coarse particles are the largest aerosols present in the atmosphere and they are formed by mechanical disintegration. The fine soil and sand particles become airborne by the wind. Turbulent air motion picks up the particles from the rough terrain where irregular soil and sand particles are present and deposits them in the atmosphere.

Types of Aerosols

The composition of different natural and anthropogenic aerosols varies primarily according to their nature of generation. Aerosols generally consist of sulfates, nitrates, sea salt, mineral dust, organics, carbonaceous components (often called soot), and so on. The primary aerosols found in the atmosphere are:

Mineral Dust Aerosols

Mineral dust aerosols are mainly generated over arid/ semi-arid regions, and deserts are significant dust sources. Dust is a mixture of quartz and clay minerals. Dust aerosols are produced due to the action of wind and weathering of soil (Gillete, 1974, 1978; Jaenicke, 1980, 1993; Prospero et al., 1983, 2002; d'Almeida, 1986; Pye, 1987; Tegen and Fung, 1994; Schwartz et al., 1995; Zender et al., 2003; Ginoux et al., 2004; Miller et al., 2004; Tegen et al., 2004).

Sea Salt Aerosols

Sea salt aerosols are the most omnipresent aerosols over the ocean and contribute about 70% to total natural aerosols in the atmosphere (Blanchard and Woodcock, 1980; Winter and Chylek, 1997; Haywood et al., 1999; Randles et al., 2004). The concentration of sea salt strongly depends on wind speed at the sea surface, and several investigators discussed its mechanism of generation due to wind (Stuhlman, 1932; Kohler, 1936; 1941; Blanchard, 1963; Fitzgerald, 1991).

Volcanic Aerosols

Volcanic eruptions are significant sources of sulfate aerosols, ash, and gaseous pollutants in the atmosphere and are often injected into the stratosphere due to strong vertical motion. The lifetime of stratospheric aerosols is longer owing to slow removal processes in the stratosphere.

Carbonaceous Aerosols

Significant sources of carbonaceous aerosols in the atmosphere are biomass burning, fossil fuel burning, outdoor burning of crop residues, forests, savannas, and gas to particle conversion of biogenic and anthropogenic volatile gases. Black and organic carbon are two broad classifications of carbonaceous aerosols (Cachier, 1998).

Organic Carbon

Organic aerosols are produced mainly by gas-to-particle conversion processes. Organic aerosols, unlike the other species, are a collective term that refers to many individual compounds. In biomass burning, organic carbon aerosols are primarily emitted, hence dominating carbonaceous aerosols emissions.

Black Carbon

Black carbon aerosols are significant outcomes in biomass and fossil fuel burning. Their absorbing nature and efficiency are decided based on graphite carbon composition, which depends on the amount of combustion (Schwartz et al., 1995). The single scattering albedo of black carbon ranges from 0.02 to 0.7, i.e., from completely absorbing aerosols to partially absorbing.

Sulfate and Nitrate Aerosols

Anthropogenic activities inject sulfate and nitrate aerosol into the atmosphere (Charlson et al., 1991). Besides anthropogenic activities, these aerosols are also generated by natural sources (i.e., volcanic eruption, bacterial action in soils, and biomass burning) (Andreae, 1985). The precursor gases for sulfate have natural as well as anthropogenic sources. Anthropogenic activities produce nitrogen-bearing gases. Nitrate aerosols also have anthropogenic and natural sources. The primary nitrogen compound in the atmosphere is nitrous oxide (N_2O) which decomposes into nitrogen and nitric oxide (NO). Vehicular and industrial emissions and combustion are the primary anthropogenic sources for nitrate aerosols in the atmosphere and fertilizers, fixation by lightning, bacterial action in soil, and biomass burning.

Non-Sea Salt Aerosols

The primary non-sea-salt sulfate (NSS) aerosol source is the gas-to-particle conversion of sulfur-bearing gases. The sulfur compounds present over the remote oceans can be of marine or continental origin. Most of the marine phytoplankton species release dimethyl sulfide (DMS), which gets oxidized by different radicals (like OH, in the presence of solar UV) to form SO_2 (Charlson et al., 1987; Clarke, 1993; Lawrence, 1993; Pandis et al., 1994; Russell et al., 1994).

Biogenic Aerosols, i.e., Bioaerosols

Aerosols injected into the atmosphere from plants and animals are called biogenic aerosols. These aerosols cover seeds, pollen, spores, bronchoscopes, and fragments of animals and plants, which vary from 1 μm to 250 μm in diameters; and bacteria, algae, viruses, protozoa, and fungi with diameters <1 μm.

These are some primary aerosols found in the atmosphere. For various reasons, most of the aerosols listed above are much more explored and studied than bioaerosols. Bioaerosols are of more concern in health-related issues than radiative forcing issues related to aerosols. The evidence and studies about bioaerosols are also spares compared to their counterparts. Here in the present chapter, there is an attempt to explore maximum information regarding bioaerosols, their types, health issues, and their role in various diseases. The chapter is structured in the following manner: First, we will introduce the bioaerosols, their classification, transport mechanism, measurement techniques, health effects, and role in the spread of the pandemic.

Introduction to Bioaerosols

Bioaerosols are a collection of various living and nonliving biological constituents. Bioaerosols are often of microbial, plant, or animal origin hosted on dust aerosols and are sometimes also known as organic dust. Fungi, bacteria (live or dead), viruses, pollen, plant residuals such as fibers, allergens, mycotoxins, and bacterial endotoxins are major bioaerosols found in the environment. Bioaerosols are abundant in indoor and outdoor environments and emitted from various natural and anthropogenic sources (Stetzenbach, 2009). Figure 1 gives a pictorial representation of the various processes associated with bioaerosols' generation, transformation, and removal. Like multiple natural aerosols, the wind is also one of the generating mechanisms for bioaerosols. The rain, wave splash, irrigation processes, wastewater treatment, harvesting, industrial and manufacturing processes, air conditioning, heating, cleaning, and various sprays are also responsible for generating bioaerosols in the indoor and outdoor environment.

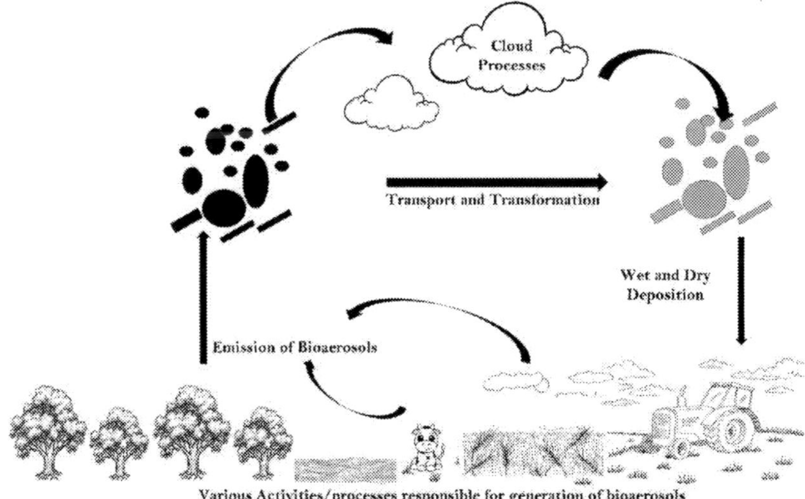

Figure 1. Schematic diagram representing the various processes associated with bioaerosols generation and disposal.

The size of bioaerosols varies over a wide range from a few nanometers to micrometers (Figure 2). Viruses are nanometer-size; bacteria are slightly bigger than viruses, about 1 µm, while fungi are greater than 1 µm. Several researchers have studied their morphological micrographs through SEM/TEM images, and Figure 3 shows some of the SEM/TEM images referred from these studies (Moon et al., 2008; Rob et al., 2018; Islam et al., 2019, Crandall et al., 2020, Li et al., 2020a).

The dust aerosols present in the atmosphere are composed of inorganic and organic materials depending on their place of origin. The micro-organism flourishes on the surface of dust in the presence of a sufficient amount of moisture. Various cultivation activities showed the most common microbe are Penicillium Aspergillus, Cladosporium, and numerous others. The indoor concentration of the bioaerosols depends on multiple factors, including the indoor environment, season, moisture, ventilation, and several others. Various studies mentioned that the farming places in the tropical region are favorite places for the growth of microbes because of the temperate climate and higher dust microscopic diversity.

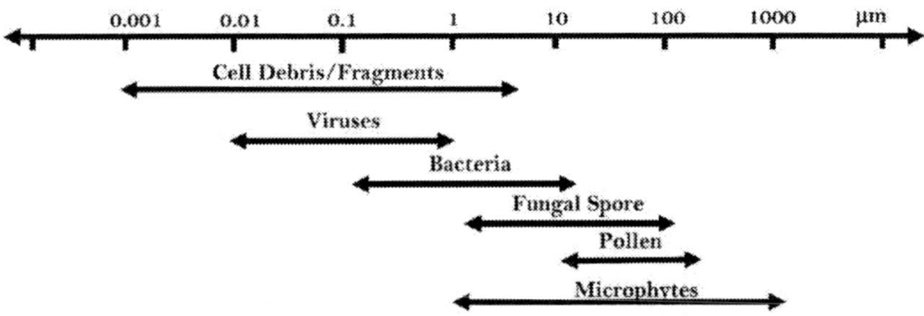

Figure 2. Schematic diagram representing the sizes of bioaerosols.

Figure 3. Morphological micrographs of bioaerosols (a) Fungi, (b) Pollen (c) Bacteria (d) Fiber (Moon et al, 2008; Rob et al., 2018; Crandall et al., 2020).

The increased concentration of bioaerosols in the indoor and outdoor environment increased the interest in bioaerosols and their impact on humans and society. Bioaerosols have a wide range of adverse health effects, thus a matter of concern. Bioaerosols are significant polluters in various industries; therefore, extensive studies are necessary concerning industrial hygiene.

Along with other emitters, industries (such as detergent, food, and waste management industries) are also potential bioaerosols emitters (Douwers et al., 2000; Wouters et al., 2002). The workers of these industries are exposed to its severe concentration, and respiratory disease symptoms are noticed (Sigsgaard et al., 1994; Schweigert et al., 2000). Several enzymes are emitted from these industries and are allergens that cause various allergic diseases. Studies have noticed a clear correlation between these enzymes' concentration and increased cases of multiple allergies (Culliana et al., 2000). The indoor environment, i.e., ventilation in the houses, emission of air pollutants from various household devices, and paints, also causes significant exposure to bioaerosols and results in various building-related diseases. Multiple studies have noticed that bioaerosols' concentration is much more prominent in indoor environments than in ambient environments (Brągoszewska et al., 2020; Li et al., 2020b). Augustynska and Posniak, 2016, have suggested the permissible limits of the different bioaerosols in the environment, as shown in Table 1 (Karol, 2020). Table 2 shows the concentration of bioaerosols found in the various studies performed at different locations.

Table 1. Proposed acceptable concentration of micro-organisms in the air (reference: Augustynska and Posniak, 2016; Karol, 2020)

Microbiological Factor	**Permissible Concentration**	
	Working spaces contaminated by organic dust	**Living quarters and public buildings**
Mesophilic bacteria	1.0×10^5 cfu m-3*	5.0×10^3 cfu m^{-3}
Gram-negative bacteria	2.0×10^4 cfu m-3*	2.0×10^2 cfu m^{-3}
Thermophilic actinobacteria	2.0×10^4 cfu m-3*	2.0×10^2 cfu m^{-3}
Fungi	5.0×10^4 cfu m-3*	5.0×10^3 cfu m^{-3}
Bacterial endotoxin	200 ng m^{-3} (2000 EU m^{-3})	5 ng m^{-3} (50 EU m^{-3})

Cfu:colony-forming units, EU: endotoxic unit, *For respirable fraction, the proposed values should be halved

Several studies have shown the severe consequence of bioaerosol exposure on human health, but its exact role in developing and enhancing disease symptoms is still unclear. There is no clear dose-response regarding the various bioaerosols, and in the lack of a robust relationship, it is challenging to quantify the bioaerosol's exact impact on human health (Douwes et al., 2003).

Table 2. Concentration of bioaerosols in air samples (Kim et al., 2018)

Site Location	Bacteria (CFU/m^3)	Fungi (CFU/m^3)	Endotoxin (ng/m^3)	1,3)-β-D-glucan (ng/m^3)	Reference
Middle East	1016 to 1973	Not done (ND)	ND	ND	Niazi et al., 2015
USA	19 to 607	3 to 59	ND	ND	McGill et al., 2015
California, USA	67 to 206	ND	ND	ND	Tsai and Macher, 2005
USA	46 to 663	ND	ND	ND	Lindemann et al., 1982
South Korea	270 to 1800	11 to 220	ND	ND	Park et al., 2013
China	209 to 838	ND	ND	ND	Guan et al., 2015
Ohio, USA	ND	ND	ND	0.81–1.2	Crawford et al., 2009
Illinois, USA	ND	ND	98 to 23,157	2.4 to 538	Yang et al., 2013
Tianjin, Northeast China	2847	ND	ND	ND	Li et al., 2020b
Amman, Jordan	132.86	372.20	ND	ND	Hussein et al., 2018
Southern Poland	~1000	ND	ND	ND	Brągoszewska et al., 2020

Classification of Bioaerosols

The bioaerosols are constituted of various components, including living and non-living organisms. Depending on the capability to reproduce, bioaerosols can be broadly classified as viable and non-viable aerosols. Viable aerosols are living organisms capable of reproduction under favorable environmental conditions. Non-viable bioaerosols are non-living organisms and thus do not reproduce. Fungi, bacteria, viruses, and pollens are prime bioaerosols concerning health issues. The concentration of these aerosols is highest in the lowest layer of the atmosphere and decreases with height. Various meteorological parameters may govern the growth and dispersion of these aerosols. These factors include temperature, humidity, wind, climatic conditions, ultraviolet radiation, and favorable surfaces such as dust aerosols. The bacteria dominate the marine bioaerosols, though the terrestrial bioaerosols are composed of bacteria, fungi, and pollen. A brief discussion regarding the prime bioaerosols is as follows:

Bacteria

Bacteria inters into the atmosphere from various surfaces by the wind. Bacteria are the prime component of the air ecosystem as they complete their entire generation cycle in a few days and significantly change their abundance in the environment. The theory of particle resuspension processes can explain the release of bacteria in the atmosphere from plants and soil with experimental evidence (Jones and Harrison, 2004). The bacterial fluxes from soil and vegetation can be monitored to assess their concentration. The upward bacterial fluxes are also correlated with sensible heat flux (Lighthart and Shaffer, 1994; Burrows et al., 2009). Researchers noticed a positive correlation between the landcover and bacterial atmospheric concentration (Bertolini et al., 2013; Smets et al., 2016). Source tracking studies also help in estimating the relative contribution of a particular source in the total concentration of bacterial load over a location. As a large portion of the earth is covered with water bodies thus, water surfaces are also a potential emitter of bacteria to the air ecosystem. Airborne bacteria can survive as individuals in the atmosphere and on pre-existing surfaces such as dust, leaf fragments, and micro-organisms. Various factors such as weather, location, time, and atmospheric composition can influence the airborne bacteria's concentration and community structure. Precipitation, temperature, air humidity, UV index, wind speed, and direction are the viral meteorological parameters that affect the ambient bacterial load (Bowers et al., 2013).

Fungi

Fungi contribute significantly to the total concentration of bioaerosols (Yeo et al., 2002; Wu et al., 2004). Usually, fungal cells are inactive when they travel in the environment but increase their concentration under high humid conditions over a wide temperature range. A few fungi cells grow their number in low moist conditions as well. Ambient concentration of fungi bioaerosols increases during the monsoon season due to increased moisture in the season (Kang et al., 2015). Researchers have investigated the outdoor fungal growth associated with meteorological parameters (Niazi et al., 2015). Damped locations, i.e., showerheads, bathrooms, toilets, damped wall and ceiling, air conditioners, air coolers, and drip pans, are prime sites for fungal growth. Exposure to UV radiation is favorable for fungal cells; some resilient ones show their survival. Fungal cells are primarily aerobic, though a few anaerobic fungi are also noticed in the environment (Haitjema et al., 2014). Various fungi grow on organic surfaces like paper, wood, textiles, and damp.

Workers involved in bio-disintegration industries are at a high risk of exposure to these microbials, leading to various health issues (Nucci and Anaissie, 2007; Velegraki et al., 2015). Fungal bioaerosols association with human health and other living organisms attracted researchers' attention, and studies are performed to explore its impact on human health and the ecosystem (Jung et al., 2009). There are multiple challenges in sampling fungi aerosols due to their biochemical nature. Studies are performed to quantify the losses while sampling and assess more accurate exposure to bioaerosols (Hamza et al., 2018). The few fungi bioaerosols are Aspergillus spp., Fusarium spp., Scedosporium spp., and Mucorales spp. Fungi disperse themselves in the environment through two mechanisms: owing to self-released energy or driven under external agencies such as wind current, gravity, or rain. Among all these, fungi bioaerosols are predominantly spread by the wind and are resistant to environmental stresses.

Pollen

Pollens are natural bioaerosols that are emitted by various flowers, trees, and grasses. These bioaerosols are usually larger and heavier than other bioaerosols and vary between 10-100 μm in general. Owing to its bigger size, these aerosols are less explored with climatic aspects. Wozniak (2016) studied the climatic impact of these aerosols and stated that coarse pollen could be ruptured into fine ones due to turbulent atmospheric conditions. Thus pollen aerosols, found in coarse and fine modes, may impact the climate. Taylor et al., 2002, performed direct observations on grass pollen release and its response to asthmatic attacks. They investigated the links between pollens and asthma.

Viruses

Viruses are also important bioaerosols and can travel a considerable distance due to their small size (Kudryashova et al., 2021). Simulation studies have shown that if fungal spores and viruses are emitted simultaneously, spores travel a few hundred meters. At the same time, viruses can travel thousands of km (Núñez et al., 2016). Viruses are noncellular units composed of DNA or RNA, but not both. They do not have their own biosynthesis system, thus requiring the living cell to replicate themselves. Once viruses reach into the respiratory system, they host on a living cell of the system. Some viruses affect the other organs and their function also. Flu viruses (A, B, and C), SARS-CoV (Severe Acute Respiratory Syndrome), SARS-CoV-2, Norwalk-like viruses, Hantaviruses, measles virus, and Varicella virus are the common viruses that are observed in the environment and cause severe illness in humans.

The flu viruses (A, B, and C types) cause influenza, a contagious respiratory illness. In the winter season, widespread of these viruses is observed. Influenza A viruses are carried by many animals, while humans have the influenza B virus. The influenza C type is less contagious than its other two counterparts and shows mild symptoms of illness. In general avian influenza, A viruses do not infect humans, but since 1997, several strains of avian influenza A have been reported in humans. The most common avian influenza A strains observed in humans are H5N2, H7N2, H9N2, and H7N3. Avian influenza is commonly known as bird flu. In humans, a new respiratory disease, severe acute respiratory syndrome (SARS), was detected at the beginning of 21 century. This viral respiratory disease is caused by SARS-associated coronavirus (SARS-CoV). Droplet transmission and direct contact with infected surfaces cause the spread of SARS. In 2019, cases of the new coronavirus, Severe Acute Respiratory Syndrome Coronavirus 2 (SARS-CoV-2), were noticed in China, and the whole world was quickly captured with this virus (Chen et al., 2020; Gorbalenya, 2020). This virus is transmitted from human to human through many pathways, i.e., by droplets, aerosols, and fomites (WHO, 2020a; Wang and Du, 2020). Severe acute respiratory illness, fever, cough, myalgia, lack of taste and smell, and weakness are prime symptoms of the *Coronavirus* disease 2019 (COVID-19) disease (Jayaweera et al., 2020; Huang et al., 2020; Judson and Munster, 2019; Nicas et al., 2005, WHO2020b). To date, approximately 248 million people got infected with this virus which caused the death of 5 million humans.

The inhalation of infected droplets also spreads the disease measles. Rubeola is also the name for the measles virus and infects humans only. In this disease, symptoms of rashes and Koplik' spots are observed along with respiratory illness symptoms. Cases of measles are noticed worldwide, but developed countries have controlled it with extensive vaccination in childhood.

Endotoxins

Endotoxins are lipopolysaccharide (LPS) and a kind of pyrogen. It has very high pro-inflammatory properties and is a part of the exterior wall of Gram-negative bacteria. A few examples of endotoxins are Escherichia coli, Salmonella, Shigella, Pseudomonas, Neisseria, Haemophilus influenza, Bordetella pertussis, and Vibrio cholera. Lung diseases, fever, blood leukocytosis, arthralgia, dyspnea, chest congestion, bronchial obstruction, and organic dust toxic syndrome result from exposure to endotoxins (Thilsing et al., 2015; Park et al., 2015; Kim et al., 2018). Endotoxins get released in small amounts during the growth of Gram-negative bacteria.

Transport

For all aerosols, including bioaerosols, the concentration over any location is governed by the place's sources, sink, and meteorology. Bioaerosols are primarily ejected into the atmosphere due to wind turbulence, and various factors govern their transport, such as meteorological conditions and characteristics of bioaerosols. Bioaerosols are mainly found in the lower atmosphere, i.e., in the planetary boundary layer. Sometimes, they can be found in the upper portion of the troposphere or the stratosphere. The wind is the main driving force for the transport of bioaerosols in the atmosphere. These can be transported on the short scale of a few millimeters to long-range, i.e., hundreds of kilometers. The short-scale transport of it occurs through clouds and long-scale transport through dust plumes. Various mechanisms are involved in the transport of the bioaerosols, and multiple tools, including modeling ones, are used to study and understand the mechanism of transport of bioaerosols.

Bioaerosols have a wide size range starting from a few nanometers to hundreds of micrometers. Along with variable sizes, their physical, chemical, and biological properties also vary greatly. Irrespective of the diversity in their properties, the motion of bioaerosols is governed by a set of forces such as buoyancy, gravity, aerodynamics lift, drag, and sometimes electrostatics forces. Mean and turbulent velocity/fluctuations, temperature and pressure gradient, and physical properties govern the strength of various forces.

Short Range Transport

The highest concentration of bioaerosols is in the lower portion of the planetary boundary layer. Wind turbulence controls the pickup of bioaerosols from the ground and their vertical mixing in the lower atmosphere. Bioaerosols can get introduced into clouds from the atmosphere and travels from one location to another, though these cannot cross a considerable distance as the cloud will rain off. These aerosols can act as cloud condensation nuclei for cloud droplet and ice crystal formation.

Long-Range Transport

The bioaerosols are transported over a considerable distance, and traces of intercontinental transport are also observed. During large-distance transport, bioaerosols are present sufficiently for a reasonably long time in the environment and impact the ecosystem, human health, crop yield, and climate (Uetake et al., 2019; Mu et al., 2020). The long-scale transport is generally with dust plumes and is transported from one continent to another. The dust from north Africa is also observed over America and sometimes north of Europe. The dust from the Gobi and Taklamakan deserts is traced over North America. Similarly, dust from Australia is also observed in New Zealand. These are examples of the long transport of dust and bioaerosols hosted on the dust.

Modeling Tools

The transport of airborne particles can be predicted over various scales with the atmospheric wind currents simulation. To simulate wind flow, mass, and energy transport, different numerical model approaches such as Gaussian, Lagrangian, Eulerian, and large eddy models are used. These modeling approaches are also used to simulate the dilution and dispersion of airborne particles after emission and/or deposition fluxes.

Measurements of Bioaerosols

Measurement of viable and non-viable microbial is essential for indoor and outdoor environments. As aforementioned, the bioaerosols' sampling data are not as robust as those available for other aerosols. Thus more efficient instruments are required for the sampling of these aerosols. Sampling of the bioaerosols is necessary to quantify the lower limit, which could lead to various health issues. A good sampler should have very high inlet efficiency. The sampler should be efficient in collecting a variety of bioaerosols with different physical properties. The sampler should be able to maintain viability and counts of bioaerosols. These properties are also called inlet efficiency, physical collection efficiency, and biological collection efficiency. These requirements should be fulfilled for good sampling so that good samples can be collected. Good inlet efficiency and physical collection efficiency are essential requirements for aerosols of any type, but biological collection efficiency is related explicitly to bioaerosols monitoring devices.

According to the bioaerosols type, the measurable properties change to assess their health impact. As for non-viable aerosols, mass and number are significant quantifiable properties. For an allergic reaction in a person, information regarding the threshold mass of the microbial is essential. In the case of viable micro-organisms, the number is vital information. The physical capture of bioaerosols, and cultured in favorable conditions is essential to know the viable counts. Bacteria and fungi cells make colonies on their host cells, while viruses make plagues. The viable count must be preserved to enhance the high biological collection efficiency and represent absolute levels of exposure. To safeguard the microbial counts in the sampler, the collecting medium should be mild to the bioaerosols, and it should be protected to get exposed from any harsh chemicals or disinfectants. The standard techniques used for sampling bioaerosols are impinger, impactors, and filtration.

Impactor

In the impactor, the sample gets collected on a solid or semi-solid collector using the properties of bioaerosols inertia. After the collection of bioaerosols, they are cultivated to know the viable counts. As physical property inertia is used to understand the viable counts, this sampler can collect and study an extensive range of bioaerosols. An airstream is flowing from the sampler, and as the property of inertia, the particles in the airstream will follow the straight-line path and impact the surface when airflow bends sharply.

Impingers

Liquid impingers also use inertia and diffusion properties to collect the microbial physically. The collecting medium in these samplers is liquid in place of a solid or semi-solid collecting medium. The impingers are specially designed bubble tubes containing a specified liquid. A known volume of air is blowing through the sampler liquid in the impinger sampler, which will react chemically or just dissolve the bioaerosols of interest.

Filtration

Filtration is a beneficial technique to separate various aerosols, including bioaerosols. This sampling technique is less efficient than others because it has very high physical collection efficiency but low biological collection efficiency. This technique is helpful for a wide range of airborne particle size materials for mass and chemical composition analysis. Still, segregating bioaerosols through the culture process may be challenging, thus reducing the techniques' biological efficiency. Filtration is an efficient technique for the purification of the air.

Bioaerosols Health Effects

Exposure to bioaerosols has numerous adverse effects on human health. Their exposure causes infectious and allergic diseases, cancer, and several other respiratory diseases. Respiratory diseases are the most common diseases owing to bioaerosols; thus, the role of bioaerosols in respiratory and allergic diseases is well studied. Various bioaerosols are found as a complex mixture of microbial and allergens. Exposure to these aerosols causes various respiratory, and infectious diseases, cancer, dermatitis, premature births, late abortions, and hormonal disbalances. Though numerous diseases are caused due to exposure to a high concentration of bioaerosols, there is a lack of sufficient evidence to establish a relation between these.

The complex mixture of various bio-aerosols causes complex health effects such as infectious diseases, allergic diseases, cancer, and toxic effects. The medical aspects of exposure to bioaerosols related to the respiratory system are well addressed compared to other health aspects.

Respiratory Diseases

Respiratory diseases are extensively explored among all the health issues caused by bioaerosols. Human lungs directly interact and are impacted by the air inheld; thus, polluted air significantly affects lung functioning. The respiratory health symptoms caused due to these organic dust ranges from acute to severe chronic diseases. The acute and mild symptoms do not need extensive attention, though comprehensive care may be required for severe symptoms. Exposure to toxins, allergens, or pro-inflammatory agents causes inflammations in the respiratory system resulting in respiratory illness. Excess exposure to toxins, allergens, and pro-inflammatory agents results in airway inflammation and respiratory illness symptoms. Various studies are performed to study the impact of organic dust on the trigger, enhancement, or development of asthma in humans, especially in children (Maheswaran et al., 2014; Lin et al., 2012; Ma et al., 2015; Karvonen et al., 2015). Studies have shown that β-glucan exposure in children below ten years causes the problem of atopic and non-atopic asthma, while it causes bronchial hyper-responsiveness in teenage children (Maheswaran et al., 2014). Mold sensitivity is reported as one of the prime causes of children's asthma intensifications (Ma et al., 2015). Researchers have noticed that indoor air abundant with mold is positively associated with asthma cases in children (Karvonen et al., 2015).

Excess exposure to pollen results in decreased lung functioning with enhancement in pulmonary inflammation (Baldacci et al., 2015). The literature says that exposure to pollen causes asthma, and grass pollen's coexistence enhances asthma risk by about 15% (Canova et al., 2013; Baldacci et al., 2015). The endotoxins and glucans significantly cause airway irritation and inflammation from bacteria and molds (Hoppin et al., 2014). Inhalation of a small number of endotoxins (~80 mg in healthy and ~20 mg in asthmatic persons) significantly decreases lung activity (Kharitonov and Sjöbring, 2007).

Infectious Diseases

Infectious diseases occur and spread due to direct and indirect contact with allergic bioaerosols consisting of bacteria, fungi, viruses, endotoxins, and pollen. The infections are transmitted directly by licking, touching, and eating and indirectly by airborne and vector transmission (Kim et al., 2018). Studies showed that persons related to veterinary services are likely to get exposed to various zoonotic infections such as Q-fever, brucellosis, anthrax, and avian and swine influenza (Wu et al., 2015; Kim et al., 2018). Pet animals and birds are also sources of various infectious diseases as they host many bacteria, viruses, and other allergens. Chlamydophila psittaci bacteria cause infection due to inhalation or exposure to infected birds like parrots and pigeons (Vanrompay et al., 2007; West, 2011; Ling et al., 2015). An outburst of Q-fever was noticed in Netherland due to airborne transmission of microbial Coxiella burnetii from goat farms, and thousands of patients were reported due to the same (Roest et al., 2011, Commandeur et al., 2014). Brooke et al. (2013) studied this microbial with the help of a human dose-response model and reported it as extremely infectious in their study. Tuberculosis (TB) is another prime contagious disease caused due to airborne bacteria which reach the lung when Mycobacterium tuberculosis is inhaled (Kim et al., 2018). The risk of spread of TB is high in dirty, low-ventilated regions, slums, and closed areas. The population living in the congregate regions and the health workers are at increased risk of getting infected with TB than in well-ventilated areas due to exposure to contaminated air (Baussano et al., 2011). Rapid growth in the bacteria on the membranes of the respiratory system causes severe infections. Various cases and deaths are reported of another infectious disease, 'Pneumonic plague', and occur when infected droplets are inhaled (Butler, 2013; Hammamieh et al., 2016). The exposure to infected animals transports the Anthrax disease in humans, which infects the intestines, lungs, or skin caused by the Bacillus anthracis (Navdarashvili et al., 2015; Azarkar and Bidaki, 2016).

Measles infections are triggered by the rubeola virus when infected droplets contact healthy humans (Clemmons et al., 2015). Reports say that millions of people got infected due to this disease and tolled millions of lives (CDA, 2015). Sometimes pneumonia, blindness, and brain damage also happen with Measles and make it lethal (Zachariah and Stockwell, 2016). Several viruses (i.e., rhinovirus, human coronavirus, respiratory syncytial virus (RSV), and adenoviruses), are responsible for common fever in humans (Allan and Arroll, 2014). These viruses can spread through various pathways, such as airborne, direct contact with contaminated objects, or from infectious secretion (sweat, urine, or nasal) (Arroll, 2011)

Cancer

Numerous factors lead to cancer, and exposure to some bioaerosols is one of them. However, there is a lack of the exact relationship of the impact of exposure to these aerosols. Lung, liver, lip, stomach, connecting tissues, biliary tract, salivary gland, and prostate cancers are cancers caused due to excess exposure to bioaerosols (Olsen et al., 1988; Blair et al., 1992; Khuder et al., 1998; Reif et al., 1989; Bethwaite et al., 2001; Johnson and Choi, 2012). Studies are performed to find possible associations between exposure to organic dust with some specific cancers ((Hayleeyesus et al., 2015; Johnson and Choi, 2012). Mycotoxins, emitted in various industries, are non-viral bioaerosols for which clear evidence exists for them to be carcinogens. Aflatoxin and ochratoxin-A are examples of mycotoxins whose excessive exposure may be a carcinogen.

Ingestion and inhalation of contagious air is the most common route of exposure to these microbials in various industries such as peanut processing, livestock food processing, and industries with grain dust (Sorenson et al., 1984; Autrup et al., 1993). In their study, Olsen et al., 1988, found that the workers at livestock processing units are at high risk of liver, biliary tract, and salivary gland. Literature has also reported that formers are at increased risk of various types of cancer due to excessive exposure to pesticides, oncogenic viruses, and other biological toxin agents associated with farm animals (Reif et al., 1989; Blair et al., 1992; Khuder et al., 1998). Association between wood dust and some specific cancer are also noticed; thus, the person associated with wood cutting, sawmills, carpentry, and joinery are at risk of cancers, particularly sinonasal cancer (Demers and Boffetta, 1998). The workers of the meat and poetry industries are at risk of lung cancer due to exposure to the high concentration of faecal material, dander, feather, and various other micro-organisms (Johnson and Choi, 2012). A high concentration of bioaerosols in meat industries is caused by animal blood, urine, faeces, and skin. A dose-response study shows a significant contribution of these in lung cancer in workers and is at about 30% more risk of lung cancer (Mclean et al., 2004).

Role of Bioaerosols in the Spread of Pandemic

In spreading infectious diseases to inhabitants, indoor air quality plays an important role. Bioaerosols, responsible for the spread of pandemics, have direct and indirect pathways. Kudryashova et al., 2021, showed through simulations that infectious aerosols propagate considerable distances in the indoor environment and can bend around obstacles.

Bioterror incident in 2001 in America with airborne Bacillus anthracis spores made it essential to study the capabilities of bioaerosols in spreading the diseases. Bioaerosols got more attention due to the pandemic outbreak of flu caused by H1N1 influenza-A virus in 2009 (Lee, 2011). The possibility of spreading pandemics through bioaerosols has strengthened with the widespread of the COVID-19 pandemic.

The World Health Organization (WHO) has also acknowledged that the SARS-CoV-2 transmission is primarily through these pathways: inhalation of infectious aerosols/droplets, deposition of droplets onto faces, hands and/or through touching an infected surface. Respiratory activities such as breathing, sneezing, talking, or coughing emits infected droplets into the environment, though the infected bioaerosols emitted in these activities may vary. COVID-19 is a new disease, and there is a lack of pervasive research, though existing analyses show that aerosols are significant routes for transmission of SARS-CoV-2. Preliminary studies (Richaed, 2020; Sia, 2020; Bao, 2020) showed the spread of SARS-CoV-2 with aerosols is relatively strong though this study was performed on animals.

Various studies have shown strong evidence that ventilation is crucial in spreading multiple infectious diseases like SARS, measles, TB, pox, and influenza (Li, 2005; 2011). Seasonal acute respiratory diseases increase in poorly ventilated places (Zhang, 2020). Ventilation is found to be an essential factor in various coronavirus diseases as well.

In the case of SARS-CoV-2, droplet and aerosol transmission are critical pathways for spreading this virus. WHO emphasizes the need for ventilation to control the COVID-19 disease, indicating that aerosols are a prime pathway in its transmission. In case of transmission through the droplet, ventilation will not make a drastic change, as droplets fall quickly on the surface under gravity and are thus unaffected by the ventilation. Ventilation makes a significant difference in the concentration of aerosols, and substantial evidence exists that this disease is spread through aerosols.

The spread of COVID-19 is more likely in a closed composite. The viral load makes a difference in the indoor and outdoor environment. In an indoor environment, the viral load could be high in case of an emitter source and poor ventilation. A case study in Japan has shown that the risk of infections increases multiple times indoors than outdoors (Nishiura, 2020). But the spread of infections indoors and in the outdoor environment is essential only in case of the spread of disease through aerosols. In droplet transmission, there is usually no difference in the case of outdoor or indoor settings. Figure 4 provides a schematic diagram for bioaerosols classification, their transport, measurement techniques, and their role in human health and spreading pandemics.

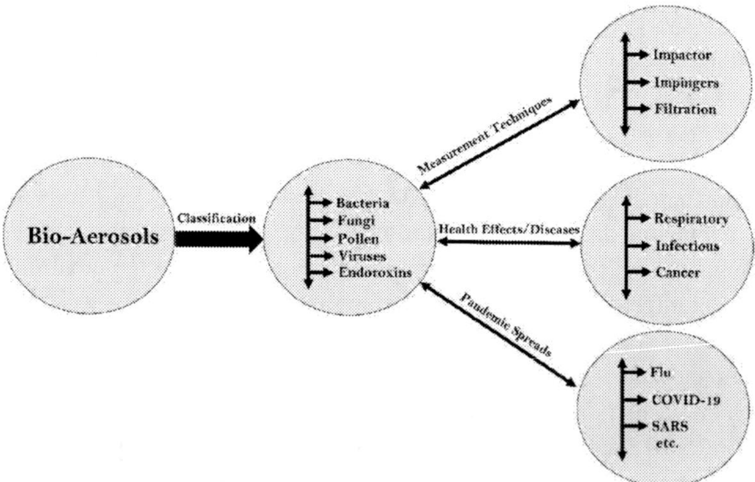

Figure 4. Schematic diagram of Bioaerosols classifications, measurement techniques, and role in health effects.

Conclusion

The present chapter explores the bioaerosol's characteristics, classifications, behavior, distribution, and role in the moderation of human health and their possible role as agents in transmitting pandemics. The bioaerosols are very much significant in both indoor and outdoor environments. Industries are also a substantial emitter of bioaerosols and cause various diseases in their workers. Bacteria, fungi, and viruses are prime bioaerosols involved in multiple respiratory, infectious diseases, and cancers. Bioaerosols are also a potential driver for spreading pandemics, especially in H1N1 and COVID-19.

References

Allan, G. M., and Arroll, B. (2014). Prevention and treatment of the common cold: making sense of the evidence. *CMAJ*, 186 (3), 190–199.

Andreae, M. O. (1985). Dimethyl sulphide in the marine atmosphere. Journal of *Geophysical Research*, *90*, 12, 891–12,900.

Arimoto, R. (2001). Eolian dust and climate: relationships to sources, tropospheric chemistry, transport and deposition. *Earth-Science Reviews*, *54*, 1–3, 29–42.

Arroll, B. (2011). Common Cold. *Clinical evidence*, 3 p. 1510 PMC, 3275147. PMID, 21406124).

Autrup, J. L., Schmidt, J., and Autrup H. (1993). Exposure to aflatoxin B1 in animal-feed production plant workers. *Environ Health Perspect*, *99*, 195–7.

Augustyńska, D., and Pośniak, M. (2016). Harmful factors in the working environment—limit values. Interdepartmental Commission for Maximum Admissible Concentrations and Intensities for Agents Harmful to Health in the Working Environment: CIOP-PIB (in Polish).

Azarkar, Z., and Bidaki, M. Z. (2016). A case report of inhalation anthrax acquired naturally. *BMC Res*. Notes, 9, 141–147.

Baldacci, S., Maio, S., Cerrai, S., Sarno, G., Baïz, N., Simoni, M., et al., (2015). Allergy and asthma: effects of the exposure to particulate matter and biological allergens. *Respir. Med*., *109*(9), 1089–1104.

Bao, L., Gao, H., Deng, W., et al., (2020). Transmission of severe acute respiratory syndrome coronavirus 2 via close contact and respiratory droplets among human angiotensin- converting enzyme 2 mice. J. *Infect. Dis*., *222*, 551–555.

Barbara J. Finlayson-Pitts, and James N. Pitts. (2000). CHAPTER 11 - Analytical Methods and Typical Atmospheric Concentrations for Gases and Particles, Editor(s): Barbara J. Finlayson-Pitts, James N. Pitts, *Chemistry of the Upper and Lower Atmosphere*. Academic Press, Pages 547-656, ISBN 9780122570605, https://doi.org/10.1016/B978-012257060-5/50013-7.

Baussano, I., Nunn, P., Williams, B., Pivetta, E., Bugiani, M., and Scano, F. (2011). Tuberculosis among health care workers. *Emerg. Infect*. Dis., *17* (3), 488–494. https://doi.org/10.3201/ei d1703.100947.

Bethwaite, P., McLean, D., Kennedy, J., and Pearce, N. (2001). Adultonset acute leukaemia and employment in the meat industry: a New Zealand case-control study. *Cancer Causes Control*, *12*, 635–43.

Bertolini, V., Gandolfi, I., Ambrosini, R., Bestetti, G., Innocente, E., Rampazzo, G., and Franzetti, A. (2013). Temporal variability and effect of environmental variables on airborne bacterial communities in an urban area of Northern Italy. *Appl. Microbiol. Biotechnol*., *97* (14), 6561e6570.

Blair, A., Hoar, Zahm S., Pearce, N. E., Heineman, E. F., and Fraumeni, J. F. (1992). Clues to cancer etiology from studies of farmers. *Scand J Work Environ Health*, *18*, 209–15.

Blanchard, D. C. (1963). The electrification of the atmosphere by particles from bubbles in the sea. *Progress in Oceanography*, vol. *1*. Pergamon, New York, pp. 71–2002.

Blanchard, D. C. and Wookcock, A. H., (1980). The production, concentration and vertical distribution of the sea-salt aerosols. *Annals of the NewYork Academy of Sciences*, *338*, 330–347.

Bowers, R. M., Clements, N., Emerson, J. B., Wiedinmyer, C., Hannigan, M. P., Fierer, N. (2013). Seasonal variability in bacterial and fungal diversity of the near-surface atmosphere. *Environ. Sci. Technol.*, *47* (21), 12097e12106.

Brągoszewska, E., Palmowska, A., and Biedroń, I. (2020). Investigation of indoor air quality in the ventilated ice rink arena. *Atmos Pollut Res*, *11*, 903–908. https://doi.org/10.1016/j.apr.2020.02.002.

Brooke, R. J., Kretzschmar, M., Mutters, N. T., and Teunis, P. (2013). Human dose response relation for airborne exposure to Coxiella burnetii. *BMC Infect. Dis.*, *13* (1), 488–495.

Burrows, S., Elbert, W., Lawrence, M., and P€oschl, U. (2009). Bacteria in the global atmosphere Part 1: review and synthesis of literature data for different ecosystems. *Atmos. Chem. Phys.*, *9* (23), 9263e9280.

Butler, T., (2013). Plague gives surprises in the first decade of the 21st century in the United States and worldwide. *Am. J. Trop. Med. Hyg.*, *89* (4), 788–793.

Cachier, H. (1998). Carbonaceous combustion aerosols. In: Harrison, R.M., Van Grieken, G.R. (Eds.), *Atmospheric Particles*, 295–348.

Canova, C., Heinrich, J., Anto, J. M., Leynaert, B., Smith, M., Kuenzli, N., Zock, J. P., et al., (2013). The influence of sensitisation to pollens and moulds on seasonal variations in asthma attacks. *Eur. Respir. J.*, *42*, 935–945.

CDC. (2015). CDC Health Advisory: U.S. Multi-State Measles Outbreak, December 2014–January 2015. US Department of Health and Human Services, CDC, Atlanta, GA (Available at http://emergency.cdc.gov/han/han00376.asp).

Charlson, R. J., Lovelock, J. E., Andreae, M. O., and Warren, S. G. (1987). Oceanic phytoplankton, atmospheric sulfur, cloud albedo and climate. *Nature*, *326*, 655–661.

Charlson, R. J., Langner, J., Rodhe, H., Leovy, C. B., and Warren, S. G. (1991). Perturbation of the Northern Hemisphere radiative balance by backscattering from anthropogenic sulphate aerosols. *Tellus*, *43AB*, 152–163.

Chen, H., Guo, J., Wang, C., Luo, F., Yu, X., Zhang, W., Li, J., Zhao, D., Xu, D., and Gong, Q. (2020). Clinical characteristics and intrauterine vertical transmission potential of COVID-19 infection in nine pregnant women: a retrospective review of medical records. *Lancet*, 395, 809–815. https://doi.org/10.1016/s0140-6736(20)30360-3.

Clarke, A. D., (1993). Atmospheric nuclei in the Pacific Midtroposphere: their nature, concentration and evolution. *Journal of Geophysical Research*, *98*, 20, 633–20, 647.

Clemmons, N. S., Gastanaduy, P. A., Fiebelkorn, A. P., Redd, S. B., and Wallace, G. S. (2015). Measles—United States. *MMWR*, *64*, 373–376.

Commandeur, M., Jeurissen, L., van der Hoek, W., Roest, H. J., and Hermans, T. C., (2014). Spatial relationships in the Q fever outbreaks 2007–2010 in the Netherlands. *Int. J. Environ. Health Res.*, *24* (2), 137–157.

Crandall Sharifa G., Norah Saarman, and Gregory S. Gilbert, (2020). Fungal spore diversity, community structure, and traits across a vegetation mosaic. *Fungal Ecology*, *45*, 100920.

Crawford, C., Reponen, T., Lee, T., Iossifova, Y., Levin, L., Adhikari, A., et al., (2009). Temporal and spatial variation of indoor and outdoor airborne fungal spores, pollen, and (1 → 3)-beta-d-glucan. *Aerobiologia*, *25* (3), 147–158.

Cullinan, P., Harris, J. M., Newman Taylor, A. J. et al., (2000). An outbreak of asthma in a modern detergent factory. Lancet, 356, 1899–900.

Demers P. A., and Boffetta P. (1998). Cancer risk from occupational exposure to wood dust. IARC Technical Report no. 32. Lyon: IARC.

Douwes, J., Thorne, P., Pearce, N., and Heederik, D. (2003). Bioaerosol Health Effects and Exposure Assessment: Progress and Prospects. *Ann. occup. Hyg*., Vol. *47*, No. 3, pp. 187–200.

Douwes, J., Dubbeld, H., van Zwieten, L. et al., (2000). Upper airway inflammation assessed by nasal lavage in compost workers: a relation with bio-aerosol exposure. *Am J Ind Med*, *37*, 459–69.

D'Almeida, G. A. (1986). A model for Saharan dust transport. *Journal of Climate and Applied Meteorology*, *25*, 903–916.

Fitzgerald, J. W. (1991). Marine aerosols: a review. *Atmospheric Environment*, *25A*, 533–545.

Gillette, D. A., Blifford, I. H. and Fryrear, D. W. (1974). The Influence of Wind Velocity on the Size Distributions of Aerosols Generated by the Wind Erosion of Soils. *J. Geophys Res*., *79*, 4068-4075.

Gillette, D. A. (1977). Fine particulate emissions due to wind-erosion. *Trans. Am. Soc. Agric. Eng*., *20*, 890.

Ginoux, P., Prospero, J. M., Torres, O., and Chin, M. (2004). Long-term simulation of dust distribution with the GOCART model: correlation with the North Atlantic oscillation. *Environmental Modeling and Software*, *19*, 113–128.

Gong, S. L., Zhang, X. Y., Zhao, T. L., McKendry, I. G., Jaffe, D. A., and Lu, N. M., (2003). Characterization of soil dust aerosol in China and its transport and distribution during 2001 ACE-Asia. 2: Model simulation and validation. *Journal of Geophysical Research*, *108* (D9).

Gorbalenya, A. E. (2020). Severe acute respiratory syndrome-related coronavirus–the species and its viruses, a statement of the coronavirus study group. *BioRxiv*, 1–15. https://doi.org/10.1101/2020.02.07.937862.

Guan, D., Guo, C., Li, Y., Lv, H., and Yu, X. (2015). Study on the concentration and distribution of the airborne bacteria in indoor air in the lecture theatres at Tianjin Chengjian University, *China. Procedia Eng*., *121*, 33–36.

Hammamieh, R., Muhie, S., Borschel, R., Gautam, A., Miller, S. A., Chakraborty, N., et al., (2016). Temporal progression of pneumonic plague in blood of nonhuman primate: a transcriptomic analysis. *PLoS One*, *11*, (3), e0151788.

Hamza Mbareche, Marc Veillette, Marie-Ève Dubuis, Bouchra Bakhiyi, Geneviève Marchand, Joseph Zayed, Jacques Lavoie, Guillaume J. Bilodeau and Caroline Duchaine, (2018). Fungal bioaerosols in biomethanization facilities. *Journal of the Air & Waste Management Association*, *68*, 11, 1198-1210, DOI: 10.1080/10962247.2018.1492472.

Haitjema, C. H., Solomon, K. V., Henske, J. K., Theodorou, M. K., and O'Malley, M. A. (2014). Anaerobic gut fungi: Advances in isolation, culture, and cellulolytic enzyme discovery for biofuel production. *Biotechnol. And Bioeng.*, *111*, 1471–1482. doi:10.1002/bit.25264.

Hayleeyesus, S. F., Ejeso, A., and Derseh, F. A. (2015). Quantitative assessment of bio-aerosols contamination in indoor air of university dormitory rooms. *Int. J. Health Sci.*, (*Qassim*), *9* (3), 249–256.

Haywood, J. M., Ramaswamy, V., and Soden, B. J. (1999). Tropospheric aerosol climate forcing in clear-sky satellite observations over the oceans. *Science*, *283*, 1299–1303.

Hoppin, J. A., Umbach, D. M., Long, S., Rinsky, J. L., Henneberger, P. K., Salo, P. M., et al., (2014). Respiratory disease in United States farmers. *Occup. Environ. Med.*, *71* (7), 484–491.

Huang, C., Wang, Y., Li, X., Ren, L., Zhao, J., Hu, Y., Zhang, L., Fan, G., Xu, J., and Gu, X. (2020). Clinical features of patients infected with 2019 novel coronavirus in Wuhan, China. *Lancet*, *395*, 497–506. https://doi.org/10.1016/s0140-6736(20)30183-5.

Hussein Tareq, Hassan Juwhari, Mustafa Al Kuisi, Hamza Alkattan, Bashar Lahlouh, and Afnan Al-Hunaiti, (2018). Accumulation and coarse mode aerosol concentrations and carbonaceous contents in the urban background atmosphere in Amman, Jordan. Arabian Journal of Geosciences, 11: 617, https://doi.org/10.1007/s12517-018-3970-z.

Husar, R. B., Prospero, J. M. and Stowe, L. L. (1997). Characterization of tropospheric aerosols over the oceans with the NOAA advanced very high resolution radiometer optical thickness operational product. *J. Geophys. Res.*, *102*, 16889-16909.

Islam, N., Dihingia, A., Khare, P., and Saikia, B. K. (2019). Atmospheric particulate matters in an Indian urban area: health implications from potentially hazardous elements, cytotoxicity, and genotoxicity studies. *Journal of Hazardous Materials*, 121472.

Jaenicke, R. (1980). Natural aerosols. *Annuals of the New York Academy of Science*, 317–329.

Jaenicke, R. (1984). Physical aspects of atmospheric aerosol, Aerosols and their climate effects. (Eds. H.E. Gerbard and A. Deepak), A Deepak Publishing, 7-34.

Jayaweera Mahesh, Hasini Perera, Buddhika Gunawardana, Jagath Manatunge, (2020). Transmission of COVID-19 virus by droplets and aerosols: A critical review on the unresolved dichotomy. *Environmental Research, 188*, 109819, https://doi.org/10.1016/j.envres.2020.109819.

Johnson, E. S., and Choi, K. M. (2012). Lung cancer risk in workers in the meat and poultry industries—a review. *Zoonoses Public Health*, *59* (5), 303–313.

Jones, A. M., and Harrison, R. M. (2004). The effects of meteorological factors on atmospheric bioaerosol concentrations: a review. Sci. *Total Environ*., *326* (1), 151e180.

Jung Jae Hee, Jung Eun Lee, Chang Ho Lee, Sang Soo Kim, and Byung Uk Lee. (2009). Treatment of Fungal Bioaerosols by a High-Temperature, Short-Time Process in a Continuous-Flow System. *Applied And Environmental Microbiology*, Vol. 75, No. 9, p. 2742–2749, doi:10.1128/AEM.01790-08.

Judson, S. D., and Munster, V. J. (2019). Nosocomial transmission of emerging viruses via aerosol-generating medical procedures. *Viruses*, *11*, 940. https://doi.org/10.3390/v11100940.

Junge, C. E. (1963). Air Chemistry and Radioactivity, Academic Press, New York.

Kang, S. M., Heo, K. J., and Lee, B. U. (2015). Why does rain increase the concentrations of environmental bioaerosols during monsoon? *Aerosol Air Qual. Res.*, *15*, 2320–2324.

Karol Bulski, 2020, Bioaerosols at plants processing materials of plant origin—a review. *Environmental Science and Pollution Research*, *27*, 27507–27514, https://doi.org/10.1007/s11356-020-09121-4.

Karvonen, A. M., Hyvärinen, A., Korppi, M., Haverinen-Shaughnessy, U., Renz, H., Pfefferle, P. I., Reme, S., Genuneit, J., and Pekkanen, J. (2015). Moisture damage and asthma: a birth cohort study. *Pediatrics*, *135* (3), 598–606.

Kaufman, Y. J., Tanré, D., and Boucher, O. (2002). A satellite view of aerosols in the climate system. *Nature*, *419*, 215–223.

Kaufman, Y. J., Boucher, O., Tanré, D., Chin, M., Remer, L. A., and Takemura, T. (2005). Aerosol anthropogenic component estimated from satellite data. *Geophys. Res. Lett.*, *32*, L17804, doi:10.1029/2005GL023125.

Khuder, S. A., Mutgi, A. B., and Schaub, E. A. (1998). Meta-analyses of brain cancer and farming. *Am J Ind Med*, *34*, 252–60.

Kharitonov, S. A., and Sjöbring, U. (2007). Lipopolysaccharide challenge of humans as a model for chronic obstructive lung disease exacerbations. Contrib. *Microbiol.*, *14*, 83–100.

Khuder S. A., Mutgi A. B., and Schaub E. A. (1998). Meta-analyses of brain cancer and farming. *Am J Ind Med*, *34*, 252–60.

Kim Ki-Hyun, Ehsanul Kabir, and Shamin Ara Jahan, (2018). Airborne bioaerosols and their impact on human health, *Journal of Environmental Sciences*, *67*, 23–35, https://doi.org/10.1016/j.jes.2017.08.027.

Kohler, H. (1936). The nucleus in and the growth of hygroscope droplets. *Transactions of the Faraday Society*, *32*, 1152–1161.

Kohler, H. (1941). An experimental investigation on sea water nuclei. Nova Acta Regional Society, *Upsaliensis*, *4*, 1–55.

Kudryashova, E., Zani, A., Vilmen, G., Sharma, A., Lu, W., Yount, J. S., Kudryashov, D. S. (2021). Inhibition of SARS-CoV-2 Infection by Human Defensin HNP1 and Retrocyclin RC-101. *J Mol Biol.*, 167225. doi: 10.1016/j.jmb.2021.167225. Epub ahead of print. PMID: 34487793; PMCID: PMC8413479.

Lawrence, M. G. (1993). An empirical analysis of the strength of the phytoplankton–dimethylsulfide–cloud–climate feedback cycle. *Journal of Geophysical Research*, *98*, 20,663–20, 673.

Lee Byung, Uk, (2011). Life Comes from the Air: A Short Review on Bioaerosol Control. *Aerosol and Air Quality Research*, *11*, 921–927, doi: 10.4209/aaqr.2011.06.0081

Ling, Y., Chen, H., Chen, X., Yang, X., Yang, J., Bavoil, P. M., et al., (2015). Epidemiology of Chlamydia psittaci infection in racing pigeons and pigeon fanciers in Beijing, China. *Zoonoses Public Health*, *62* (5), 401–406.

Lin, S., Jones, R., Munsie, J. P., Nayak, S. G., Fitzgerald, E. F., and Hwang, S. A. (2012). Childhood asthma and indoor allergen exposure and sensitization in Buffalo. New York. *Int. J. Hyg. Environ. Health*, *215* (3), 297–305.

Li, Y., Duan, S., Yu, I. T., Wong, T. W. (2005). Multi-zone modeling of probable SARS virus transmission by airflow between flats in Block E, Amoy Gardens. *Indoor Air*, *15*, 96–111.

Li, M., Qi, J., Zhang, H., Huang, S., Li, L., and Gao, D. (2011). Concentration and size distribution of bioaerosols in an outdoor environment in the Qingdao coastal region. Sci. *Total Environ*., *409*, 3812–3819, doi:10.1016/j.scitotenv.2011.06.001.

Li, W., Liu, L., Xu, L., Zhang, J., Yuan, Q., Ding, X., et al., (2020a). Overview of primary biological aerosol particles from a Chinese boreal forest: Insight into morphology, size, and mixing state at microscopic scale. *Science of The Total Environment*, 137520.

Li, Y., Ge, Y., Wu, C., Guan, D., Liu, J., and Wang, F. (2020b). Assessment of culturable airborne bacteria of indoor environments in classrooms, dormitories and dining hall at university: a case study in China. *Aerobiologia*. https://doi.org/10.1007/s10453-020-09633-z.

Lighthart, B., and Shaffer, B. (1994). Bacterial flux from chaparral into the atmosphere in mid-summer at a high desert location. *Atmos. Environ*., *28* (7), 1267e1274.

Lindemann, J., Constantinidou, H. A., Barchet, W. R., and Upper, C. D. (1982). Plants as sources of airborne bacteria, including ice nucleation-active bacteria. *Appl. Environ. Microbiol*., 44, 1059–1063.

IPCC, Climate Change 2013, (2013). The Physical Science Basis, Working Group I Contribution to the Fifth Assessment Report of the Intergovernmental Panel on Climate Change.

Ma, Y., Tian, G., Tang, F., Yu, B., Chen, Y., Cui, Y., He, Q., and Gao, Z. (2015). The link between mold sensitivity and asthma severity in a cohort of northern Chinese patients. *J. Thorac. Dis*., (4), 585–590.

Maheswaran, D., Zeng, Y., Chan-Yeung, M., Scott, J., Osornio-Vargas, A., Becker, A. B., et al., (2014). Exposure to beta- (1,3)-D-glucan in house dust at age 7–10 is associated with airway hyperresponsiveness and atopic asthma by age 11–14. *PLoS One*, *9* (6), e98878.

Maring, H., Savoie, D. L., Izaguirre, M. A., Custals, L., and Reid, J. S. (2003). Mineral dust aerosol size distribution change during atmospheric transport. *Journal of Geophysical Research*, *108* (D19), 8592.

McClatchey, R. A., Fenn, R. W., Selby, J. E. A., Volz, F. E. and Garing, J. S. (1972). Optical properties of the atmosphere. 3rd ed. *AFCRL Environ. Res*. Papers No. 411, 108 pp.

McGill, G., Moore, J., Sharpe, T., Downey, D., and Oyedele, L. (2015). Airborne Bacteria and Fungi Concentrations in Airtight Contemporary Dwellings. Mackintosh School of Architecture > Mackintosh Environmental Architecture Research Unit (MEARU).

McLean, D., Cheng, S., Woodward, A., and Pearce, N., (2004). Mortality and cancer incidence in New Zealand meat workers. *Occup. Environ. Med*., *61* (6), 541–547. https://doi.org/10.1136/oem.2003.010587.

Miller, R. L., Tegen, I., and Perlwitz, J. (2004). Surface radiative forcing by soil dust aerosols and the hydrologic cycle. *Journal of Geophysical Research– Atmospheres*, *109* (D4) Art. No. D04203.

Mishchenko, M. I., Travis, L. D., Kahn, R. A., and West, R. A. (1997). Modeling phase functions for dustlike tropospheric aerosols using a shape mixture of randomly oriented polydisperse spheroids, *J. Geophys. Res.*, *102*(D14), 16,831–16,847.

Moon, Hye-Kyoung, Stefan Vinckier, Erik F Smets, S. Huysmans, (2008). Palynological evolutionary trends within the tribe Mentheae with special emphasis on subtribe Menthinae (Nepetoideae: Lamiaceae), *Plant Systematics and Evolution*, *275*(1), 93-108, DOI:10.1007/s00606-008-0042-y.

Mu, F. F., Li, Y. P., Lu, R., Qi, Y. Z., Xie, W. W., and Bai, W. Y. (2020). Source identification of airborne bacteria in the mountainous area and the urban areas. *Atmospheric Research*, *231*, 104676.

Navdarashvili, A., Doker, T. J., Geleishvili, M., Haberling, D. L., Kharod, G. A., and Rush, T. H., (2015). Human anthrax outbreak associated with livestock exposure Georgia, 2012. *Epidemiol. Infect*. *19*, 1–12.

Nicas, M., Nazaroff, W. W., and Hubbard, A., (2005). Toward understanding the risk of secondary airborne infection: emission of respirable pathogens. *J. Occup. Environ. Hyg.*, *2*, 143–154. https://doi.org/10.1080/15459620590918466.

Niazi, S., Hassanvand, M. S., Mahvi, A. H., Nabizadeh, R., Alimohammadi, M., Nabavi, S., et al., (2015). Assessment of bioaerosol contamination (bacteria and fungi) in the largest urban wastewater treatment plant in the Middle East. *Environ. Sci. Pollut. Res.*, *22* (20), 16014–16021.

Nishiura, H., Oshitani, H., Kobayashi, T., Saito, T., Sunagawa, T., Matsui, T., et al., (2020). Closed environments facilitate secondary transmission of coronavirus disease 2019 (COVID-19). medRxiv.

Nucci, M., and E. Anaissie, (2007). Fusarium infections in immunocomprised patients. *Clin. Microbiol. Rev*., 20 (4), 695–704. doi:10.1128/CMR.00014-07.

Núñez, A., Amo de Paz, G., Rastrojo, A., García Ruiz, A. M., Alcamí, A., Gutiérrez-Bustillo, A. M., and Moreno Gómez, D. A. (2016). Monitoring of airborne biological particles in outdoor atmosphere. Part 1: importance, variability and ratios, *Int. Microbiol*., *19*, pp. 1-13.

Olsen, J. H., Dragsted, L., and Autrup, H. (1988). Cancer risk and occupational exposure to aflatoxins in Denmark. *Br J Cancer*, *58*, 392–6.

Park, D. U., Yeom, J. K., Lee, W. J. and Lee, K. M., (2013). Assessment of the levels of airborne bacteria, gram-negative bacteria, and fungi in hospital lobbies. *Int. J. Environ. Res. Public Health*, *10* (2), 541–555. https://doi.org/10.3390/ijerph10020541.

Park, S. M., Kwak, Y. S., and Ji, J. G. (2015). The effects of combined exercise on health-related fitness, endotoxin, and immune function of postmenopausal women with abdominal obesity, *J. Immunol. Res*., 830567.

Pandis, S. N., Russell, L. M., and Seinfeld, J. H. (1994). The relationship between DMS flux and CCN concentration in remote marine regions. *Journal of Geophysical Research*, *99*, 16,945–16, 957.

Prospero, J. M., Charlson, R. J., Mohnen, B., Jaenicke, R., Delany, A. C., Mayers, J., Zoller, W. and Rahn, K. (1983). The atmospheric aerosol system - An overview, *Rev. Geophys Space Phys*., *21*, 1607-1629.

Prospero, J. M., et al., (2002). Environmental characterization of global sources **of** atmospheric soil dust identified with the Nimbus 7 Total Ozone Mapping Spectrometer (TOMS) absorbing aerosol product. *Reviews of Geophysics*, (1), Art. No. 1002.

Pye, K. (1987). Aeolian Dust and Dust Deposits. Academic Press, London 334p.

Randles, C. A., Russell, L. M. and Ramaswamy, V. (2004). Hygroscopic and optical properties of organic sea salt aerosol and consequences for climate forcing, *Geophys. Res. Lett*, Vol. *31*, L16108.

Reist, P. C. (1984). Introduction to Aerosol science, MacMillan Publication company.

Reif, J. S., Pearce, N. E., and Fraser, J. (1989). Cancer risks among New Zealand meat workers. *Scand J Work Environ Health, 15*, 24–9.

Rob Cornelissen, Andreas Bøggild, Raghavendran Thiruvallur Eachambadi, Roman Koning, Anna Kremer, Silvia Hidalgo-Martinez, Eva-Maria Zetsche, Lars R. Damgaard, Robin Bonné, Jeroen Drijkoningen, Jeanine S. Geelhoed, Thomas Boesen, Henricus T. S. Boschker, Roland Valcke, Lars Peter Nielsen, Jan D'Haen, Jean V. Manca and Filip J. R. Meysman, (2018). The Cell Envelope Structure of Cable Bacteria, *Frontiers in Microbiology*, Volume *9*, Article 3044, doi: 10.3389/fmicb.2018.03044.

Roest, H. I., Tilburg, J. J. H. C., Vellema, P., Van Zijderveld, F. G., Klaassen, C. H. W., and Raoult, D., (2011). The Q fever epidemic in The Netherlands: history, onset, response and reflection. *Epidemiol. Infect. 139* (1), 1–12.

Russell, L. M., Pandis, S. N., and Seinfeld, J. H. (1994). Aerosol production and growth in marine boundary layer. *Journal of Geophysical Research*, *99*, 20, 989–21,003.

Satheesh S. K. and Krishnamoorthy, K. (2005). Radiative effects of Natural aerosols: A review, *Atmospheric Environment*, *39*, 2089-2110.

Sia, S. F., Yan, L. M., Chin, A. W. H., et al., (2020). Pathogenesis and transmission of SARSCoV-2 in golden hamsters. *Nature*, *583*, 834–838, https://doi.org/10.1038/s41586-020-2342-5

Schweigert, M. K., Mackenzie, D. P., and Sarlo, K. (2000). Occupational asthma and allergy associated with the use of enzymes in the detergent industry – a review of the epidemiology, toxicology and methods of prevention. *Clin Exp Allergy*, *30*, 1511–8.

Schwartz, S. E., et al., (1995). Group Report: Connections Between Aerosol Properties and Forcing of Climate.

Seinfeld, J. H. and Pandis, S. N. (1998). Atmospheric Chemistry and Physics: From Air Pollution to Climate Change. Wiley, NewYork.

Sigsgaard T., Malmros P., Nersting L., and Petersen C. (1994). Respiratory disorders and atopy in Danish refuse workers. *Am J Respir Crit Care Med*, *149*, 1407–12.

Smets Wenke, Serena Moretti, Siegfried Denys, and Sarah Lebeer, (2016). Airborne bacteria in the atmosphere: Presence, purpose, and potential. *Atmospheric Environment*, *139*, 214-221, http://dx.doi.org/10.1016/j.atmosenv.2016.05.038.

Sorenson W. G., Jones W., Simpson J., and Davidson J. I. (1984). Aflatoxin in respirable airborne peanut dust. *J. Toxicol Environ. Health*, *14*, 525–33.

Stetzenbach L. D. (2009). Airborne Infectious Microorganisms, *Encyclopedia of Microbiology*, 175–182. doi: 10.1016/B978-012373944-5.00177-2.

Stuhlman Jr., O. (1932). The mechanics of effervescence. *Journal of Applied Physics, 2* (6), 457–466.

Taylor, P. E., Flagan, R. C., Valenta, R., and Glovsky, M. M. (2002). Release of allergens as respirable aerosols: A link between grass pollen and asthma. *J Allergy Clin Immunol. 109*(1), 51-6. doi: 10.1067/mai.2002.120759. PMID: 11799365.

Tegen, I., Werner, M., Harrison, S. P., and Kohfeld, K. E. (2004). Relative importance of climate and land use in determining present and future global soil dust emission. *Geophysics Research Letters*, *31*, L05105.

Tegen, I. and Fung, I. (1994). Modeling of Mineral dust in the atmosphere: Source, transport and optical thickness, *J. Geophys. Res.*, *99*, 22897-22914.

Thilsing, T., Madsen, A. M., Basinas, I., Schlünssen, V., Tendal, K., and Bælum, J. (2015). Dust, endotoxin, fungi, and bacteria exposure as determined by work task, season, and type of plant in a flower greenhouse. *Ann. Occup. Hyg.*, *59* (2), 142–157.

Tsai, F. C., and Macher, J. M. (2005). Concentrations of airborne culturable bacteria in 100 large US office buildings from the BASE study. *Indoor Air*, *15*, 71–81. https://doi.org/10.1111/j.1600-0668.2005.00346.x.

Uetake, J., Tobo, Y., Uji, Y., Hill, T. C. J., Demott, P. J., Kreidenweis, S. M., and Misumi, R., (2019). Seasonal changes of airborne bacterial communities over Tokyo and influence of local meteorology. *Frontiers in Microbiology*, *10*, 1572.

Vanrompay, D., Harkinezhad, T., Walle, M., and Beeckman, D., (2007). Droogenbroeck C. Chlamydophila psittaci transmission from pet birds to humans. *Emerg. Infect. Dis.*, *13*, 1108–1110.

Velegraki, A., Cafarchia, C., Gaitanis, G., Iatta, R. and T. Boekhou. (2015). Malassezia infections in humans and animals: Pathophysiology, detection, and treatment. *PloS Pathog.*, *11* (1), e1004523. doi:10.1371/journal.ppat.1004523.

Wang, J., and Du, G., (2020). COVID-19 may transmit through aerosol. *Ir. J. Med. Sci.*, 1–2. https://doi.org/10.1007/s11845-020-02218-2.

West, A. (2011). A brief review of Chlamydophila psittaci in birds and humans. *J. Exot. Pet. Med.*, *20*, 18–20.

Winter, B. and Chylek, P. (1997). Contribution of sea salt aerosol to the planetary clearsky albedo. *Tellus Series B-Chemical and Physical Meteorology*, *49*(1), 72-79.

Wouters, I. M., Hilhorst, S. K. M., Kleppem P. et al., (2002). Upper airway inflammation and respiratory symptoms in domestic waste collectors. *Occup Environ Med*, *59*, 106–12.

Wozniak, M. C. (2016). A 10-year climatology of pollen aerosol for the continental United States: implications for aerosol-climate interactions. vol. 2016.

World Health Organization (WHO). (2020a). Coronavirus Disease (COVID-2019) Situation Reports, 05 May 2020. https://www.who.int/emergencies/diseases/novelcoronavirus- 2019/situation-reports/, Accessed date: 6 May 2020.

World Health Organization (WHO). (2020b). Modes of Transmission of Virus Causing COVID-19: Implications for IPC Precaution Recommendations: Scientific Brief, 27 March 2020. https://apps.who.int/iris/handle/10665/331601.

Wu, P. C., Tsai, J. C., Li, F. C., Lung, S. C. and Su. H. J. (2004). Increased levels of ambient fungal spores in Taiwan are associated with dust events from China. *Atmos. Environ.*, *38*, 4879–4886.

Wu, X., Lu, Y., Zhou, S., Chen, L., and Xu, B. (2015). Impact of climate change on human infectious diseases: empirical evidence and human adaptation. *Environ. Int.*, *86*, 14–23.

Yang, X., Wang, X., Zhang, Y., Lee, J., Su, J., and Gates, R. S. (2013). Monitoring total endotoxin and (1 → 3)-beta-D-glucan at the air exhaust of concentrated animal feeding operations. *J. Air Waste Manag. Assoc.*, *63* (10), 1190–1198.

Yeo, H., and J. Kim, (2002). SPM and fungal spores in the ambient air of west Korea during the Asian dust (yellow sand) period. *Atmos. Environ.*, *36*, 5437– 5442.

Zachariah, P., and Stockwell, M. S. (2016). Measles vaccine: past, present, and future. *J. Clin. Pharmacol.*, *56* (2), 133–140.

Zender, C., Bian, H., and Newman, D. (2003). Mineral dust entrainment and deposition (dead) model description and 1990s dust climatology. *Journal of Geophysical Research*, *108*, 4416.

Zhang Renyi, Yixin Li, Annie L. Zhang, Yuan Wang, and Mario J. Molina, (2020). Identifying airborne transmission as the dominant route for the spread of COVID-19. *PNAS*, vol. *117*, no. 26, 14857–14863, www.pnas.org/cgi/doi/10.1073/pnas.2009637117.

Chapter 6

Source Apportionment of Atmospheric Aerosol Using Receptor Models: A Simple to Complex Approach

S. K. Sharma*
Environmental Sciences and Biomedical Metrology Division, CSIR-National Physical Laboratory, New Delhi, India

Abstract

Thousands of articles have been published in the field of source apportionment of atmospheric aerosols ($PM_{0.1}$, PM_1, $PM_{2.5}$, PM_{10}, and TSP) in sub-urban and urban cities around the world, using various statistical tools and marker elements present in the aerosol to design mitigation strategies to improve global air quality. In this chapter, we have discussed the identifications and quantifications of possible sources of $PM_{2.5}$ at a specific location of megacity Delhi, India using various statistical tools/receptor models [simple to complex: IMPROVE protocol, Enrichment Factor (EF), Principal Component Analysis (PCA), UNMIX, Positive Matrix Factorization (PMF)]. All these models identified the soil/crustal dust (SD), industrial emissions (IE), secondary aerosols (SA), vehicular emissions (VE), biomass burning (BB), fossil fuel combustion (FFC), and sea salts (SS) are the dominant sources of $PM_{2.5}$ over the receptor site of megacity Delhi. This chapter discusses the effectiveness of various statistical methods used in source identifications and quantifications of atmospheric aerosol, which can help the stakeholders and authorities in identifying emission control initiatives to improve ambient air quality.

* Corresponding Author's Email: sudhir.npl@nic.in; sudhircsir@gmail.com.

In: Atmospheric Aerosols
Editor: Binoy K Saikia
ISBN: 979-8-88697-211-5

Keywords: aerosols, source apportionment, PCA, UNMIX, PMF

1. Introduction

The quantification of the possible sources of the atmospheric aerosol ($PM_{2.5}$, PM_{10}, TSP, etc.) for mitigation and improving urban air quality is a major issue in the present scenario (Hopke et al., 2020). Therefore, the development and application of statistical methods are needed to accurately identify and quantify aerosol sources over any region (Paatero and Tapper, 1994; Sharma et al., 2014; Hopke and Jaffe, 2020; Jain et al., 2017; 2020) around the globe. Source apportionment of aerosol is vital in identifying specific aerosol concentrations to the adverse health effects (Pope and Dockery, 2006). There are several statistical methods such that enrichment factor (EF), IMPROVE protocol, principal component analysis (PCA), UNMIX, chemical mass balance (CMB), positive matrix factorization (PMF), etc. which are applied for source apportionment analysis of pollutants (Paatero and Tapper, 1993; 1994; Tauler et al., 1993; 1994; Sharma et al., 2014). Some are obsolete (EFs, PCA, IMPROVE, etc.); however, some are still being used, and some are relevant (UNMIX, CMB and PMF). EF is a primeval quantitative data analytical method (first used in the 1960s) that explains the origin of the elements and their abundance in the ambient aerosols (Taylor and McLennan, 1995). It only provides the crude information that the element is more abundant than the average earth's crustal values and does not explain the elemental abundance (Hopke and Jaffe, 2020). Similarly, PCA provides information about the factor loading based on the eigenvector method, an elementary screening tool (Roscoe et al., 1982). Hopke and Jaffe (2020) have suggested that these tools (EF and PCA) may not be used as receptor models for source apportionment of atmospheric aerosol because more quantitative data analytical tools had been developed and available.

The source apportionment of atmospheric aerosol is essential for researchers, stakeholders, academicians, and policymakers to learn more about the characteristics of particulates, the impact of regional and local sources, as well as their influence on air quality in the region. For the last two decades, mainly two categories of statistical tools have been used as receptor models (RMs) around the world, firstly, chemical mass balance (CMB) and secondly, multivariate factor analysis models such as PCA, UNMIX, and PMF (Hopke et al., 2003; Jain et al., 2017; 2020). Even though with the same data sets and chemical species/constituents of atmospheric aerosol (e.g., PM_1,

$PM_{2.5}$, PM_{10} and TSP etc.), these receptor models resolve the different sources with different apportioned due to different statistical procedures and constraints. Therefore, applying these models together with the same input data sets would prefer comparing their results, which helps precisely identify the crucial sources. In a recent global review, Hopke et al., (2020) reported that the secondary inorganic species, sea salts, traffic, industry, biomass burning, coal/oil combustion and secondary organic aerosols are the significant sources of $PM_{2.5}$ and PM_{10}. Figure 1 shows the source profile of $PM_{2.5}$ resolved by the PMF model during 2013-2016 in Delhi, India (Jain et al., 2020).

This chapter discusses the various statistical tools/ methods/ models used in the source identification and quantification of atmospheric aerosol. The various receptor models have been applied with the same chemical species of $PM_{2.5}$ collected at an urban site of megacity Delhi during January 2013-December 2014 to evaluate the performance and relevance of these models in source apportionment study.

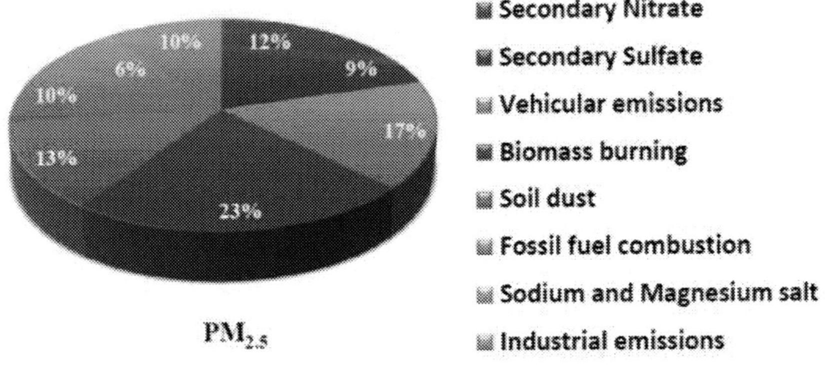

Figure 1. Source profile of $PM_{2.5}$ at Delhi using PMF model.

2. Source Apportionment of Aerosol Using Receptor Models

Fundamentally, the RMs (PCA/APCS, UNMIX and PMF) used mass conservation (chemical mass balance) of chemical species of pollutants for their source apportionment study. A Schematic flow chart for source apportionment procedure are given in Figure 2 as discussed by Belis et al., (2014). The primary assumption of the source apportionment is that the chemical species of pollutants remain unaltered throughout its respective source regions to receptor sites with the same time dependence (Belis et al., 2013). The general equations of RMs are:

$$X=GF+E \tag{1}$$

Here, all the variables represent a matrix, and they are X for the concentration of pollutant, G denotes source contribution, F refers to source profile, and E stands for unsuitable elemental concentration (Paatero and Tapper, 1994; Hopke et al., 2006). In India, PCA has been used to conduct the majority of source apportionment studies in the last two decades, but with the changing scenario, scientific communities have been showing a proclivity for employing the PMF model to quantify sources of PM due to its idiosyncrasies (Pant and Harrison, 2012; Banerjee et al., 2015; Jain et al., 2018). The application of UNMIX model for source apportionment study in India is limited, and till 2014 only 3% of studies have been reported (Banerjee et al., 2015). Since all these three models (PCA/APCS, UNMIX, and PMF) employ distinct statistical processes, thus the obtained results also vary. As a result, integrating all three models on the same data sets would result in more robust source profiles and contributions and better interpret the results.

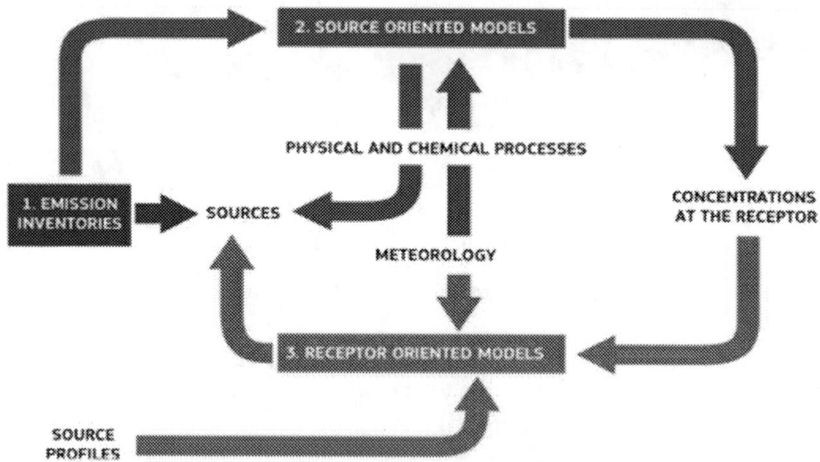

Figure 2. Schematic illustration for the procedure for source apportionment (Belis et al., 2014).

The chemical species of $PM_{2.5}$ were used as inputs of various receptor models (EF, IMPROVE protocol, PCA, UNMIX, PMF) and extracted the possible sources of $PM_{2.5}$ over Delhi. The mean concentrations of OC, EC, inorganic ions and elemental concentrations of $PM_{2.5}$ samples collected during 2013-14 are depicted in Table 1. During 2013-2014, the mean concentration of $PM_{2.5}$ was estimated at 122 ± 94 μg m^{-3}. Almost all the chemical constituents of $PM_{2.5}$ were computed maximum in the winter season, followed by summer and monsoon. The seasonal changes in the mean value of OC, EC, WSIC and elements (of $PM_{2.5}$) with ranges are given in Table 6.1.

Table 1. The mean concentrations of $PM_{2.5}$ and their chemical species (in μg m^{-3}) in Delhi

Species			Seasons		
	Mean	Range	Winter	Summer	Monsoon
$PM_{2.5}$	121 ± 93	27.2–435	215^a ± 93.3	82.9^a ± 24.7	66.1^a ± 52.2
OC	17.8 ± 14.2	3.26–69.2	31.1^a ± 15.2	11.5^a ± 3.72	9.9^a ± 8.12
EC	10.3 ± 8.05	0.87–35.8	17.8^a ± 7.78	7.16^a ± 3.05	5.57^a ± 5.42
F^-	0.92 ± 0.70	0.06–3.74	1.07^b ± 0.81	0.91^b ± 0.51	0.73^b ± 0.65
Cl^-	7.76 ± 5.71	0.90–31.4	10.8^a ± 6.65	5.65^a ± 3.02	6.47^a ± 5.12
SO_4^{2}	12.8 ± 8.09	1.98–56.3	16.8^b ± 11.1	10.4^b ± 3.84	11.4^b ± 5.12
NO_3^-	10.1 ± 9.83	0.22–52.4	18.7^a ± 11.3	5.85^a ± 2.04	4.17^a ± 3.15
NH_4^+	9.41 ± 8.57	0.16–45.5	16.3^a ± 10.3	8.35^a ± 2.95	3.44^a ± 3.74
Na^+	5.07 ± 3.10	0.95–18.8	5.11^b ± 2.84	3.85^b ± 1.71	6.08^b ± 3.83
K^+	4.11 ± 2.71	0.32–16.7	5.22^b ± 2.52	4.06^b ± 2.42	2.95^b ± 2.61
Si	2.07 ± 0.52	1.03–4.23	3.12 ± 1.31	1.93 ± 0.62	1.13 ± 0.54
Mg	0.95 ± 0.87	0.12–4.27	1.33^b ± 1.45	0.45^b ± 0.24	1.05^b ± 0.61
Ca	2.84 ± 2.23	0.36–13.8	3.41^b ± 2.75	2.98^b ± 1.71	2.06^b ± 1.59
Al	2.87 ± 0.95	0.28–5.31	3.44^b ± 0.83	2.38^b ± 0.81	2.71^b ± 1.02
S	2.82 ± 1.63	0.14–5.63	3.42^a ± 1.47	1.28^a ± 0.95	1.37^a ± 0.81
Cr	0.04 ± 0.11	0.03–0.93	0.10^b ± 0.17	0.03^b ± 0.02	0.02^b ± 0.01
Ti	0.13 ± 0.21	0.02–0.97	0.23^b ± 0.30	0.06^b ± 0.03	0.04^b ± 0.10
Fe	0.27 ± 0.42	0.09–2.04	0.51^b ± 0.48	0.12^b ± 0.16	0.07^b ± 0.12
Zn	0.14 ± 0.22	0.03–1.37	0.28^b ± 0.25	0.04^b ± 0.03	0.05^b ± 0.07
Mn	0.04 ± 0.03	0.003–0.13	0.03^b ± 0.03	0.02^b ± 0.02	0.01^b ± 0.01
Cu	0.05 ± 0.04	0.004–0.14	0.07^b ± 0.19	0.02^b ± 0.02	0.01^b ± 0.01
As	0.07 ± 0.15	0.003–0.71	0.14^b ± 0.21	0.02^b ± 0.02	0.01^b ± 0.01
Br	0.04 ± 0.10	0.004–0.42	0.07^b ± 0.15	0.02^b ± 0.01	0.04^b ± 0.05
Pb	0.03 ± 0.02	0.002–0.15	0.05^b ± 0.02	0.02^b ± 0.01	0.01^b ± 0.05

± Standard deviation; aSignificantly different ($p < 0.05$); bNot significantly different ($p > 0.05$)

3. Enrichment Factor (EF)

The enrichment factor of the elements present in $PM_{2.5}$ samples has been computed to identify the origin of the elements, i.e., anthropogenic or natural, along with their relative abundance in the ambient aerosols. The EF of the elements' concentration was calculated using equation 2 (Taylor and McLennan, 1995) and depicted in Figure 3.

$$EF = (El_{sample}/X_{sample})/(El_{crust}/X_{crust}) \quad (2)$$

Here, El_{sample} is the concentration of elements (El) in sample, X_{sample} is the concentration of reference element (X), El_{crust} is the concentration of element (El) in upper continental crust, X_{crust} is the concentration of reference element (X) in upper continental crust. Generally, Fe, Ti and Al are used as the reference element for EF calculation. In the present case, Al is used as a reference element for EF calculation, which is primarily used for the urban area of India.

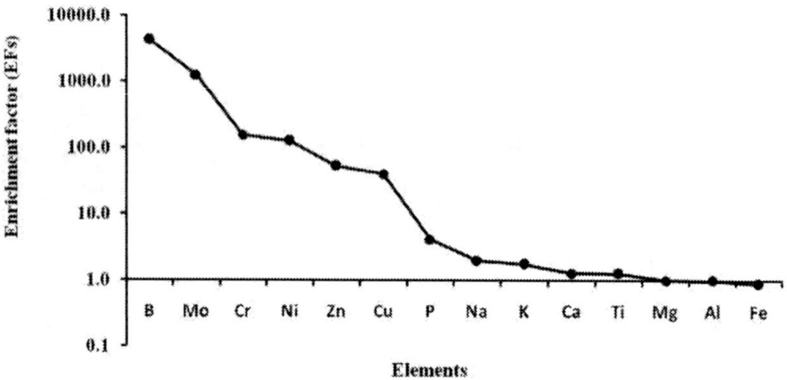

Figure 3. Enrichment factors (EFs) of elements present in $PM_{2.5}$ at Delhi.

The enrichment factor of the several elements presents in $PM_{2.5}$ has been calculated using the reference element (Figure 3). In the present case, the mean value of EFs of elements like Al, Fe, Ti, K, Mg, Ca, Na and P range from < 1 to 5, which signifies that the elements mainly arise from the soil/crustal dust. The mean EFs value of elements like Cu, Zn, Ni, and Cr range from 5 to 160, signifying that they come from the soil and non-soil emissions sources (Jain et al., 2020). The higher EFs of Cr and Zn in $PM_{2.5}$ samples also belonged from industrial sources (Shridhar et al., 2010).

4. Mass Closure (IMPROVE Model)

In this instance, the mass closure of the $PM_{2.5}$ constituents was reconstructed ($RCPM_{2.5}$) by IMPROVE equation (Chan et al., 1997; Malm et al., 2007) using the addition of concentrations of particulate organic matter (POM), ammonium sulphate (AS), sea salts (SS), ammonium nitrate (AN), light-absorbing carbon (LAC), and soil, as represented below:

$$RCPM2.5 = [POM] + [AS] + [SS] + [AN] + [LAC] + [Soil]$$

The change between the concentrations of $PM_{2.5}$ and $RCPM_{2.5}$ (dM = $PM_{2.5} - RCPM_{10}$) are said to be the mass difference (dM) or unidentified mass (UM). The chemical reactions between SO_2 and related alkaline gases (like NH_3) and particulates produce AS within the atmosphere. The average AS in $PM_{2.5}$ samples was assessed as 17.7 μg m^{-3} ($1.375[SO_4^{2-}]$), accounting for 14.5% of $RCPM_{2.5}$. In the ambient atmosphere, the neutralization reaction of gaseous NH_3 and HNO_3 produces AN. The concentration of reconstructed AN ($1.29[NO_3^-]$) was 12.9 μg m^{-3}, accounting for 10.6% of $RCPM_{10}$ (Table 2).

POM (1.6[OC]) in $PM_{2.5}$ samples was calculated as 28.6 μg m^{-3} (23.4% of $RCPM_{10}$). The main source of POM in the atmosphere are fossil fuel combustion and biomass burning, whereas the secondary POM is formed by the oxidation of gas-phase precursors in the ambient air emitted from anthropogenic and biogenic sources. LAC is mainly composed of black carbon, elemental carbon, and graphite carbon, which originates from partial combustion of fossil fuels as well as biomass combustion. In this case, the average LAC was computed as 10.4 μg m^{-3} (8.5% of $RCPM_{10}$). Some of the sources of soil dust within the atmosphere are paved and unpaved roads, fire, farming and construction activities, deserts, etc. (Seinfeld and Pandis, 1998). The average reconstructed concentrations of soil [1.16{(1.90[Al] + 2.15[Si] + 1.41[Ca] + 2.09[Fe] + 1.67[Ti])}] in Delhi was 16.9 μg m^{-3} (13.9% of $RCPM_{2.5}$). The UM between $PM_{2.5}$ and $RCPM_{10}$ was recorded as 15.8 μg m^{-3}, accounting for 13% to $PM_{2.5}$ concentrations.

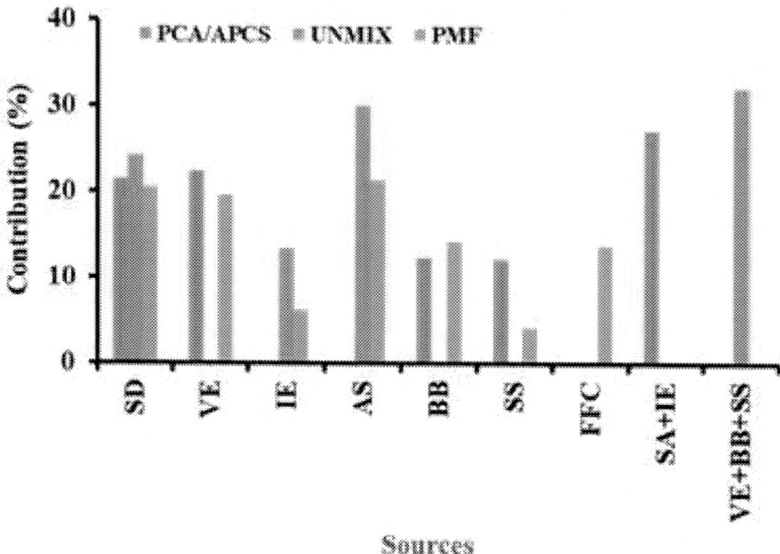

Figure 4. Percentage contributions of $PM_{2.5}$ sources resolved by three receptor models.

5. Principal Component Analysis (PCA)

Chemical constituents of $PM_{2.5}$ have been applied as input to PCA-APCS using Varimax rotation. The first factor was extracted as secondary aerosols+industrial emissions (SA + IE), elucidated about 23.4% variance, and contributed 27.2% for $PM_{2.5}$ concentrations. Ammonium, sulphate, and nitrate are used as markers for identifying SA, and the high loads of elements like As, Zn, Fe, Cu, Cr, Pb and S specify the presence of industrial emissions, which makes mixed type source (SA + IE).

For the second factor, the high concentrations of OC, EC, Zn, Fe, Mn, and Pb referred to VE accounted for 22.4% of $PM_{2.5}$ (Figure 4) explained based on 23% of the variance. In general, EC is widely used as a tracer for combustion sources (Song et al., 2006; Robles et al., 2008; Yin et al., 2010), whereas OC might derive from primary emission sources (Turpin et al., 1991; Ho et al., 2003; Behera and Sharma, 2010). The high concentration of Pb, Zn, Fe, Mn, Cu, and Al are also contributed to the transportation source.

Table 2. $PM_{2.5}$ concentrations reconstructed by IMPROVE equation

$PM_{2.5}$ (sources)	IMPROVE model	$PM_{2.5}$ ($\mu g\ m^{-3}$)	$PM_{2.5}$ (%)
AS	$1.375[SO_4^{2-}]$	17.7	14.5
AN	$1.29[NO_3^-]$	12.9	10.6
POM	1.6[OC]	28.6	23.4
LAC	LAC = [EC]	10.4	8.5
SS	$2.54[Cl^-]$	19.7	16.1
Soil	1.16{1.90[Al] + 2.15[Si] +1.41[Ca] + 2.09[Fe] + 1.67[Ti]}. Where factor 1.16 compensates for excluding MgO, Na_2O, K_2O and H_2O from the crustal mass calculation.	16.9	13.9
Reconstructed PM mass	AS+AN+POM+LAC+SS+Soil	106.2	87.0
Mass difference (dM)	PM – RCPM	15.8	13.0

AS: Ammonium sulphate; AN: Ammonium nitrate; POM: Particulate organic matter; LAC: Light absorbing carbon; SS: Sea salt.

The third factor is resolved as sea salt (with 12.2% of the variance) with the abundance of Na^+, Cl^-, Mg^{2+} and K^+ in $PM_{2.5}$ samples at Delhi (Srimuruganandam and Nagendra, 2012a; 2012b). The high loading of Al, Si, Ca, Mn, Mg and Fe in $PM_{2.5}$ concentrations at sampling site-resolved as soil/crustal dust (21.5% PM2.5) with 8.4% variance (Gupta et al., 2007; Banerjee et al., 2015). The fifth factor resolved as biomass burning, which explained 7.4% of the variance (Rodriguez et al., 2004; Andersen et al., 2007; Pant and Harrison, 2012). Our prior publication contains detailed information about the extraction of sources of $PM_{2.5}$ using PCA/APCS (Jain et al., 2020).

6. UNMIX

UNMIX model resolved the four sources (SD, IE, SA and VE+BB+SS), including a mixed source of $PM_{2.5}$ at the urban site of Delhi (Figure 4). The first source was extracted as SD (24.3% of $PM_{2.5}$ concentration) due to the high loading of Al, Ti, Fe, Ca, and Mg (Jain et al., 2017). The second source resolved as mixed type (VE+BB+SS) source with the abundance of OC, EC, Zn, Fe, K^+, Na^+, Cl^-, Ca, Mg and SO_4^{2-} which contributes about 32.2% for

$PM_{2.5}$. The third source corresponds to secondary aerosols (30% of $PM_{2.5}$ mass concentration) due to the high loading of nitrate, sulfate and ammonium as analyzed by UNMIX model. The UNMIX model resolved the fourth source as IE (13.4% for $PM_{2.5}$) due to the abundance of elements like Cr, Zn, As, Fe and Cu in $PM_{2.5}$ (Sharma et al., 2016; Jain et al., 2017).

7. Positive Matrix Factorization (PMF)

PMF model was applied on the chemical species of $PM_{2.5}$, collected at an urban site of Delhi, which identified the seven sources of $PM_{2.5}$ (SD, VE, SS, VE, SA, BB and FFC), which is depicted in Figure 5.

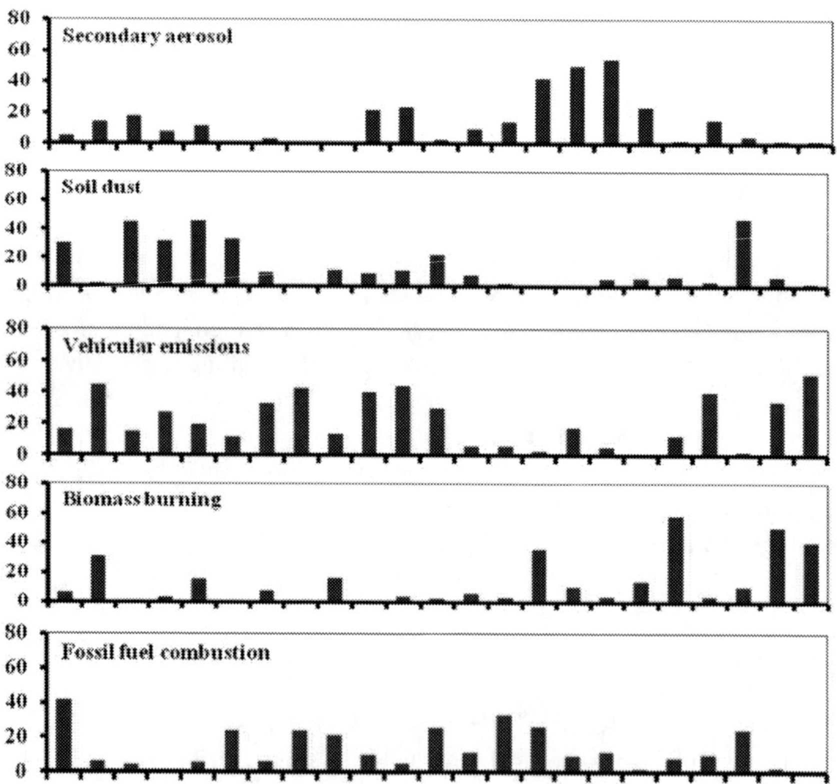

Figure 5. (Continued).

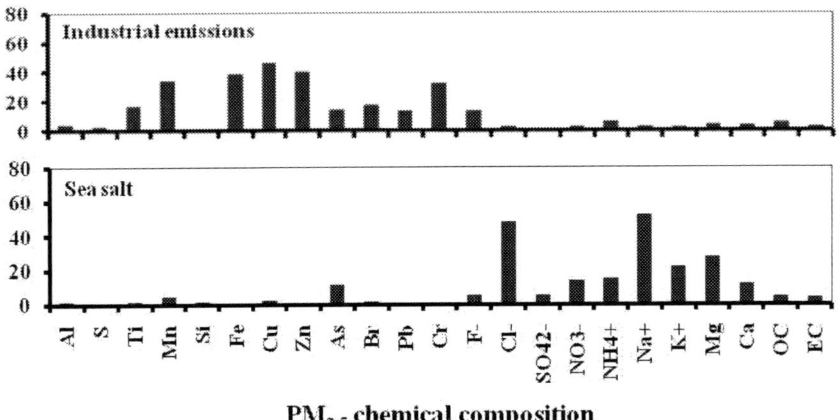

Figure 5. PMF factor profiles of $PM_{2.5}$.

Source 1: PMF extracted SA with mass loading of 21.3% of $PM_{2.5}$ by applying the key markers like NO_3^-, SO_4^{2-} and NH_4^+ (Figure 5). Source 2: Due to the high loading of Fe, Ca, Na, Mg, Al and K, PMF resolved SD as another source of $PM_{2.5}$, accounting for 20.5% of $PM_{2.5}$ mass concentration (Lough et al., 2005). Generally, Al, Si, Ca, Ti, Fe, Pb, Cu, Cr, Ni, Co and Mg are used to tracer soil/crustal dust in India (Sharma et al., 2014). Source 3: VE is contributed to 19.7% of $PM_{2.5}$ as extracted by the PMF model. Source 4: The burning of wood, biomass, and vegetation characterized the abundance of K^+ and SO_4^{2-} in aerosol samples (Wu et al., 2007). The PMF analysis indicated that BB accounts for 14.3% of $PM_{2.5}$ mass concentration. Source 5: The PMF extracted 13.7% FFC for $PM_{2.5}$ with high concentrations of Al, Cl, Fe, Zn Cr and SO_4^{2-} Source 6: The PMF analysis resolved the IE, accounting for 6.2% of $PM_{2.5}$. Source 7: A high loading of Na, K and Cl in $PM_{2.5}$ implies the influence of SS, which is accounted for 4.3% of $PM_{2.5}$ (Jain et al., 2017; 2020).

8. Comparison of Models

The present study, IMPROVE model recorded the highest contribution of POM (23.5% of $PM_{2.5}$) apart from the other sources, e.g., soil/crustal dust (13.9%), AS (14.5%), AN (10.6%), SS (16.1%) and LAC (16.1%). Apart from the IMPROVE protocol and enrichment factor analysis, the current study highlights the source analysis of $PM_{2.5}$ using three different receptor models (PCA-APCS, UNMIX, and PMF), which allows researchers to understand better the evaluation of these models concur, contrast, or complement each another. PCA/APCS model resolved five sources [SD, Mixed type (SA + IE), VE, SS, BB] of $PM_{2.5}$. However, UNMIX model resolved four sources [SD, mixed type (VE+BB+SS), SA and IE], and the PMF model resolved seven sources (SD, VE, SA, BB, SS, IE, FFC) of $PM_{2.5}$ in Delhi (Jain et al., 2017; 2020). Each model identifies different source apportionments based on the selected variables (marker elements present in $PM_{2.5}$ samples). The derived source profiles may be compared to the input matrix without transition because the PMF model uses a point-by-point least-squares reduction strategy rather than the correlation matrix information (Jain et al., 2020). In contrast, the load matrix of PCA is non-dimensional, and the source profiles are obtained using multi-linear regression (MLR) analysis. Also, PCA differs from UNMIX and PMF since it does not contribute to a non-negativity constraint (Belis et al., 2013). A comparison of the percentage contribution of major sources of aerosols over the Indian region and other countries is summarized in Table 3, representing similar types of sources.

Table 3. Average contributions (in %) of major sources of PM mass concentrations in the literature

Location	PM ($\mu g/m^3$)	Sea salt (%)	Soil dust (%)	Vehicular emissions (%)	Secondary aerosols (%)	Other sources (%)	References
Delhi, India	249.7	-	16.9	23.3	–	58.8	Jain et al., 2018
		-	11.8	18.5	27	42.7	Jain et al., 2018
		4.8	22.7	17	20.5	35.0	Jain et al., 2018
Delhi, India	177.9	4.4	20.7	17	21.7	36.2	Sharma et al., (2014b)
Delhi, India	-	-	23	14	22.7	40.3	CPCB report (2010)
Delhi, India	-	-	14.2	58.2	-	30.8	Srivastava and Jain (2007)
Delhi, India	-	-	35	62	-	3	Srivastava et al. (2008)
Delhi, India	219	-	27	-	-	45.0	Tiwari et al. (2009)
Delhi, India	161	16	37.2	23.2	2.6		Tiwari et al., (2013)
Mumbai, India	114	15	17.7	23	-	31.9	Chelani et al., (2008)
Kolkata, India	-	-	37	-	-	35.0	Gupta et al., (2007)

Location	PM ($\mu g/m^3$)	Sea salt (%)	Soil dust (%)	Vehicular emissions (%)	Secondary aerosols (%)	Other sources (%)	References
Hyderabad, India	135.1	-	40	22	-	28	Gummeneni et al., (2011)
Chennai, India	-	40.4	3.4	16	22.9	13.4	Srimuruganandam & Nagendra (2012)
Chennai, India	70-130	27.9	24.7	25.3	-	22.1	Anu et al., (2013)
Chennai, India	–	27	13	35	19	6	Selvaraju et al., (2013)
Zaragoza, Spain	–	7	40	9	-	30	Callen et al., (2009)
Shinjung, Taiwan	39.5	8.4	34	24.92	24.33	8.35	Gugamsetty et al., (2012)
Salamanca, Mexico	89.1		39.87	30.16	14.42	1.82	Murillo et al., (2012)
Brisbane, Australia	7.2	58	8	30	22	32	Chan et al., (2011)
Taiyuan, China	33.0	-	12	13	16	30	Zeng et al., (2010)
Ile-Ife, Nigeria	41.7	-	29	5	-	66	Ogundele et al., (2016)

SA: secondary aerosols; BB: biomass burning; SS: sea salt; IE: industrial emissions; FFC: fossil fuel combustion.

Conclusion

In this chapter, we have discussed the source apportionment of $PM_{2.5}$ at a specific location of Delhi, India using various statistical tools to assess the efficacy of receptor models. All these models identified the SD, SA, VE, IE, BB, FFC and SS as the probable sources of $PM_{2.5}$ at megacity Delhi. In this chapter, the effectiveness of various statistical methods used in source identifications and quantifications of atmospheric aerosols are also highlighted, which may aid in better understanding the emergence of regional and local sources of aerosols and thus enhance the ambient air quality.

Acknowledgments

The author is thankful to the Director, CSIR-NPL and Head, ES&BMD, CSIR-NPL, for their encouragement and support. The author also acknowledges the Council of Scientific & Industrial Research, India, for financial support for this study. The author thankfully acknowledges one of his Ph.D. students (Srishti Jain) for her contribution to the source apportionment study as well as the contribution of text materials of this chapter.

References

Andersen Z J, Wahlin P, Raaschou-Nielsen O, Scheike T, Loft, S (2007) Ambient particle source apportionment and daily hospital admissions among children and elderly in Copenhagen. *J. Expo Sci. Environ. Epidemiol.* 17 (7): 625-636.

Banerjee T, Murari V, Kumar M, Raju M P (2015) Source apportionment of airborne particulates through receptor modeling: Indian scenario. *Atmos. Res.* 164:167-187.

Behera S N, Sharma M (2010) Investigating the potential role of ammonia in ion chemistry of fine particulate matter formation for an urban environment. *Sci. Total Environ.* 408(17): 3569-3575.

Belis C A, Karagulian F, Larsen B R, Hopke P K (2013) Critical review and meta-analysis of ambient particulate matter source apportionment using receptor models in Europe. *Atmos. Environ.* 69: 94-108.

Chen L W A, Watson J G, Chow J C, Magliano K L (2007) Quantifying $PM_{2.5}$ source contributions for the San Joaquin Valley with multivariate receptor models. *Environ. Sci. Technol.* 41(8): 2818-2826.

Gupta A K, Karar K, Srivastava A (2007) Chemical mass balance source apportionment of PM_{10} and TSP in residential and industrial sites of an urban region of Kolkata, India. *J. Hazardous Materials* 142 (1): 279-287.

Ho K F, Lee S C, Chow J C, Watson J G (2003) Characterization of PM_{10} and $PM_{2.5}$ source profiles for fugitive dust in Hong Kong. *Atmos Environ.* 37(8):1023-1032.

Hopke P K (2003) Recent developments in receptor modeling. *J. Chemometrics* 17(5):255-265.

Hopke P K, Dai Q, Li L, Feng Y (2020) Global review of recent source apportionments for airborne particulate matter. *Sci. Tot. Environ.* 740:140091

Hopke P K, Jaffe D A (2020) Letter to the Editor: Ending the use of obsolete data analysis methods. *Aero Air Qual. Res.* 20:688-689.

Jain S, Sharma S K, Choudhary N, Masiwal R, Saxena M, Sharma A, Mandal T K, Gupta A, Gupta N C, Sharma C, (2017) Chemical characteristics and source apportionment of $PM_{2.5}$ using PCA/APCS, UNMIX and PMF at an urban site of Delhi, India. *Environ. Sci. Poll. Res.* 24: 14637-14656.

Jain S, Sharma S K, Vijyan N, Mandal T K (2020) Seasonal characteristics of aerosols ($PM_{2.5}$ and PM_{10}) and their source apportionment using PMF: A four-year study over Delhi, India. *Environ. Poll.* 262:114337.

Kar S, Maity J P, Samal A C, Santra S C (2010) Metallic components of traffic-induced urban aerosol, their spatial variation, and source apportionment. *Environ. Monit. Asses.* 168(1-4): 561-574.

Karar K, Gupta, A K (2007) Source apportionment of PM_{10} at residential and industrial sites of an urban region of Kolkata, India. *Atmos. Res.* 84(1):30-41.

Kothai P, Saradhi I V, Prathibha P, Hopke P K, Pandit G G, Puranik V D (2008) Source apportionment of coarse and fine particulate matter at Navi Mumbai, India. *Aerosol. and Air Quality Research* 8(4): 423-436.

Kulshrestha A, Satsangi P G, Masih J, Taneja A (2009) Metal concentration of $PM_{2.5}$ and PM_{10} particles and seasonal variations in urban and rural environment of Agra, India. *Sci. Total Environ.* 407(24): 6196-6204.

Kumar A V, Patil R S, Nambi K S V (2001) Source apportionment of suspended particulate matter at two traffic junctions in Mumbai, India. *Atmos. Environ.* 35(25): 4245-4251.

Lee D D, Seung H S (1999) Learning the parts of objects by non-negative matrix factorization. *Nature* 401: 788–791.

Lee J H, Hopke P K (2006) Apportioning sources of $PM_{2.5}$ in St. Louis, MO using speciation trends network data. *Atmos. Environ.* 40:360-377.

Malm W C, Pitchford M L, McDade C, Ashbaugh L L (2007) Coarse particle speciation at selected locations in the rural continental United States. *Atmos. Environ.* 41(10): 2225-2239.

Paatero P, Tapper U (1993) Analysis of different modes of factor analysis as least-squares fit problems. *Chemom. Intell. Lab. Syst.* 18: 183–194.

Paatero P, Tapper U, (1994) Positive matrix factorization: A non-negative factor model with optimal utilization of error estimates of data values. *Environmetrics* 5: 111–126.

Pant P, Harrison R M (2012) Critical review of receptor modelling for particulate matter: a case study of India. *Atmos. Environ.* 49:1-12.

Pope III C A, Dockery D W (2006) Health effects of fine particulate air pollution: lines that connect. *J. Air & Waste Management Association* 56(6): 709-742.

Rodrıguez S, Querol X, Alastuey A, Viana M M, Alarcon M, Mantilla E, Ruiz C R (2004) Comparative PM_{10}–$PM_{2.5}$ source contribution study at rural, urban and industrial sites during PM episodes in Eastern Spain. *Sci. Total Environ.* 328(1): 95-113.

Roscoe B A, Hopke P K, Dattner S L, Jenks J M (1982) The use of principal components factor analysis to interpret particulate compositional data sets. *J. Air Pollut. Control. Assoc.* 32: 637–642.

Seinfeld J H, Pandis S N (2016) *Atmospheric chemistry and physics: from air pollution to climate change.* John Wiley & Sons.

Sharma S K, Mandal T K, Jain S, Saraswati, Sharma A, Saxena M (2016) Source Apportionment of $PM_{2.5}$ in Delhi, India Using PMF Model. *Bull. Environ. Contamin. Toxicol.* 97 (2): 286-293.

Sharma S K, Mandal T K, Saxena M, Rashmi, Datta A, Saud T (2014) Variation of OC, EC, WSIC and trace metals of PM_{10} over Delhi. *J. Atmos Solar-Terres Phys.* 113: 10-22.

Sharma S K, Singh A K, Saud T, Mandal T K, Saxena M, Singh S, Raha S (2012) Study on water-soluble ionic composition of PM_{10} and related trace gases over Bay of Bengal during W_ICARB campaign. *Meteorol. Atmos. Phy.* 118(1-2):37-51.

Shridhar V, Khillare P S, Agarwal T, Ray S (2010) Metallic species in ambient particulate matter at rural and urban location of Delhi. *J. Hazardous Materials* 175(1):600-607.

SongY, Xie S, Zhang Y, Zeng L, Salmon L G, Zheng M (2006) Source apportionment of $PM_{2.5}$ in Beijing using principal component analysis/absolute principal component scores and UNMIX. *Sci. Total Environ.* 372(1):278-286.

Srimuruganandam B, Nagendra S S (2012a) Source characterization of PM_{10} and $PM_{2.5}$ mass using a chemical mass balance model at urban roadside. *Sci. Total Environ.* 433: 8-19.

Srimuruganandam B, Nagendra S S (2012b) Application of positive matrix factorization in characterization of PM_{10} and $PM_{2.5}$ emission sources at urban roadside. *Chemosphere* 88(1):120-130.

Tauler R, Kowalski B, Fleming S (1993) Multivariate curve resolution applied to spectral data from multiple runs of an industrial process. *Anal. Chem.* 65: 2040–2047.

Turpin B J, Huntzicker J J (1991) Secondary formation of organic aerosol in the Los Angeles Basin: a descriptive analysis of organic and elemental carbon concentrations. *Atmos Environ. Part A*. General Topics 25(2):207-215.

Yin J, Harrison R M, Chen Q, Rutter A, Schauer J J (2010) Source apportionment of fine particles at urban background and rural sites in the UK atmosphere. *Atmos. Environ.* 44(6): 841-851.

Chapter 7

Source Identification of Aerosols Using Stable Carbon and Nitrogen Isotopic Composition

S. K. Sharma* and T. K. Mandal
Environmental Sciences and Biomedical Metrology Division, CSIR-National Physical Laboratory, Dr. K. S. Krishnan Road, New Delhi, India

Abstract

Recently radiocarbon and nitrogen analysis of aerosols have been used as a powerful technique to identify (decoding and tracking biogeochemical processes) the sources of aerosols around the globe. To characterize the sources of aerosols over the Indian region, stable isotopic ratios of carbon ($\delta^{13}C_{TC}$) and nitrogen ($\delta^{15}N_{TN}$) along with total carbon (TC) and total nitrogen (TN) have been reported in this chapter. $\delta^{13}C_{TC}$ and $\delta^{13}T_{TN}$ analysis of aerosols ($PM_{2.5}$ and PM_{10}) over the most polluted Indo-Gangetic Plain (IGP) region of India revealed that fossil-fuel combustion (FFC), biomass / wood burning (dung cake, C3 & C4 plant matter and wood, etc.), and secondary aerosols (SAs) are the important sources of aerosols over the IGP region. Levels of $\delta^{13}C_{TC}$ and $\delta^{13}T_{TN}$ detected in aerosols over the Himalayan region of India reveal biomass burning (crop residue and wood-burning), FFC and waste incineration as the important sources of aerosols. In this chapter, we discussed the composition of $\delta^{13}C_{TC}$ and $\delta^{15}N_{TN}$ in atmospheric aerosols (i.e., $PM_{2.5}$ and PM_{10}) and their possible sources.

Keywords: $PM_{2.5}$, PM_{10}, $\delta^{13}C_{TC}$ and $\delta^{13}T_{TN}$

---

* Corresponding Author's Email: sudhir.npl@nic.in; sudhircsir@gmail.com.

In: Atmospheric Aerosols
Editor: Binoy K Saikia
ISBN: 979-8-88697-211-5

Introduction

The deterioration of air quality of an urban region is usually associated with the rise in vehicular traffic, industrial emissions (IE), biomass burning (BB), formation of secondary aerosols (SAs) through the gas to particle conversion and uncontrolled urban growth (Kim et al. 2014; Sharma et al. 2015). The air quality studies indicated that the aerosols, especially fine mode ($\leq$2.5 μm) particles, can lead to severe cardiovascular and respiratory disorders (Dockery and Pope, 1994; Pope et al. 2009). Therefore, source apportionment studies of aerosols ($PM_{2.5}$ and PM_{10}) are essential to develop air quality control strategies over a region (Waked et al., 2014). Many statistical tools have been used to resolve this issue for the source apportionment of aerosols (Paatero and Tapper, 1994; Paatero, 1997; Ulbrich et al., 2009). Aerosols comprising carbonaceous aerosols (CAs) lead to poor air quality and harmful effects on climate and atmospheric chemistry (Liousse et al. 1996; Jacobson, 2001). The major sources of CAs are wood-burning, combustion of biomass, fossil fuels and other carbon containing materials combustion (Claeys et al. 2004; Venkataraman et al. 2005; Wang et al. 2006; Rengrajan et al. 2007; Sharma et al. 2014a; Jain et al. 2016). Recently radiocarbon analysis studies have been demonstrated by several researchers and reported as a powerful technique for source identification for CAs (Chen et al. 2013; Gustafsson et al. 2009; Sharma et al. 2015; Sen et al. 2018).

Exploration of $\delta^{13}C_{TC}$ and $\delta^{15}N_{TN}$ are well recognized for decoding and tracking in oceanography and limnology (Naru et al. 2007; Peng et al. 2015) but limited in the aerosol study (Gomez et al. 2018). Court et al. (1981) were probably the first to suggest the utility of C isotopes in identifying the urban air particulate matter (PM). Widory et al. (2004) applied the isotopic characteristics of aerosols to identify the aerosol sources in an urban atmosphere. To investigate atmospheric N cycling in particulate matter, the isotopic ($\delta^{15}N$) characteristics of NH_4^+ and NO_3^- were also used by Moore (1977). Thereafter, $\delta^{15}N$ isotope has been widely used for source identification in studies around the globe (Kawamura et al. 2004; Russell et al. 1998; Turekian et al. 1998; Widory, 2007; Yeatman et al. 2001). Though, characterization of C and N isotopes for decoding the probable sources of PM over Asia is limited (Kawamura et al. 2004). Such information is also sparse over the Indian region, considered the most polluted region (Sharma et al. 2015; Sen et al. 2018). This chapter summarizes the application of stable isotopes of C and N ($\delta^{13}C$ and $\delta^{15}N$) in aerosols studies to identify PM ($PM_{2.5}$ and PM_{10}) sources over India and the globe.

Analytical Techniques

In this section, we discuss the general chemical procedures for estimating the stable C & N isotopic ratios, carbon components (OC and EC) and ionic species (NH_4^+ and NO_3^-) of $PM_{2.5}$ and PM_{10} (Figure 1). The $^{13}C/^{12}C$ and $^{15}N/^{14}N$ ratios along with TC and TN content of $PM_{2.5}$ and PM_{10} are analyzed by an isotope-ratio mass spectrometer coupled with an elemental analyzer (Sharma et al. 2015). The level of $\delta^{13}C$ and $\delta^{15}N$ corresponding to V-PDB (Vienna-Peedee Belemnite) and atmospheric N2 standards, respectively estimated as:

$$\delta^{13}C \text{ and } \delta^{15}N = ((R_{sample}-R_{standard})/R_{standard}) \times 1000 \qquad (1)$$

where, R = $^{13}C/^{12}C$ and $^{15}N/^{14}N$.

The isotopic analysis procedure, the standard used, calibration and repeatability errors are given in Sharma et al., (2015). The concentrations of OC and EC in PM ($PM_{2.5}$ and PM_{10}) samples are performed using OC/EC carbon analyzer following the IMPROVE-A protocol (Chow et al., 2004; Sharma et al., 2017). The concentrations of ionic species (NH_4^+ and NO_3^-) of $PM_{2.5}$ and PM_{10} are determined using Ion Chromatography (Sharma et al. 2017). The schematic diagram for the chemical characterization procedures for $PM_{2.5}$ and PM_{10} is depicted in Figure 1.

Carbon and Nitrogen Isotopic Composition of $PM_{2.5}$ and PM_{10}

The stable isotope ratio of carbon ($\delta^{13}C$) presents the origin, sources, and ageing consequences of carbonaceous components in particulates (Fisseha et al. 2009; Bosch et al. 2014; Bikkina et al. 2016). The $\delta^{13}C$ levels in C3 and C4 plants depend upon photosynthesis pathways for fixing the atmospheric CO_2, and its value varies and depleted (range: -20 to -32‰) (Smith and Epstein 1971; Bosch et al. 2014). Mkoma et al. (2014) reported that particulate bound $\delta^{13}C$ coming from burning of C3 plants ranged from -23‰ to -30‰, whereas C4 plants ranged from -14‰ to -20‰. Generally, $\delta^{13}C$ values in C3 plants range from -23 to -30‰, whereas C4 plants range from -12 to -20‰ (Cachier et al. 1986; Turekian et al. 1998; Martinelli et al. 2002). Widory (2006) reported coal combustion emissions have $\delta^{13}C$ of -23.6 ± 0.7‰ with a range of -22.9 to -24.9‰ whereas Agnihotri et al. (2011) analyzed average $\delta^{13}C$ levels as -22‰. Also, the $\delta^{13}C$ levels in natural gas combustion were -28.1 to -39.7‰ compared to petrol and diesel (-24 to -28‰) (Widory et al. 2004; Widory 2006; Pavuluri and Kawamura 2012; Miyazaki et al. 2012; Fisseha et al. 2009).

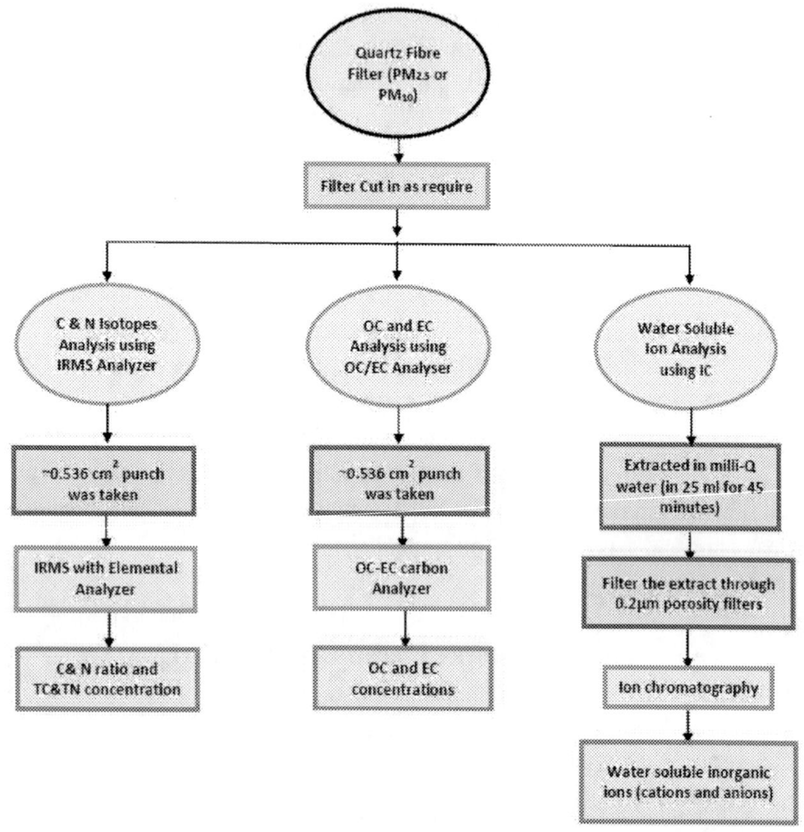

Figure 1. Schematic diagram for the chemical characterization procedures for $PM_{2.5}$ and PM_{10}.

C & N isotopic studies of aerosols conducted at different sites of India and other countries are given in Table 1. A land campaign study was conducted at Mohal-Kullu, Kolkata, Darjeeling, Lucknow and Ajmer where average $\delta^{13}C_{TC}$ were estimated as -25.9 ± 0.6‰, -25.5 ± 1.0‰, -25.6 ± 0.9‰, -24.8 ± 1.0‰ and -24.8 ± 0.5‰, respectively (Sen et al. 2018). Sharma et al. (2015) had examined the $\delta^{13}C_{TC}$ levels in PM_{10} at Delhi (-25.5 ± 0.5‰) and Kolkata (-

26.0 ± 0.4‰) based on a yearlong study during 2011 and observed almost identical levels of $\delta^{13}C_{TC}$ at both the sites. $\delta^{13}C_{TC}$ levels PM_{10} and TSP at megacities Mumbai and Chennai (situated outside IGP region of India) were reported as -25.9 ± 0.3‰ and (-25 ± 0.2‰, respectively (Aggarwal et al. 2013; Pavuluri et al. 2011). A nearly similar level of $\delta^{13}C_{TC}$ levels in PM_{10} (~ -25.6 to -25.9‰) was detected in the Himalayan region (at Mohal-Kullu, Nainital and Darjeeling) of India (Sen et al. 2018; Hegde et al. 2016). During winter, huge amounts of biomass fuels are burnt for heating purposes over IGP and the Himalayan region of India could be account for the heavy $\delta^{13}C$ levels (Bikkina et al. 2016; Sharma et al. 2015; Hegde et al. 2016).

The $\delta^{15}N$ levels ranges from -15 to 30‰ from the point sources (Agnihotri et al. 2011; Widory 2007). Turekian et al. (1998) reported a large variability in $\delta^{15}N_{TN}$ levels emitted from the combustion of both C3 and C4 plants (Table 2). The average level of $\delta^{15}N$ emitted from diesel, unleaded gasoline and coal were 4.6 ± 0.8‰, ~ -4.6‰ and ~5.3‰, respectively (Widory et al. 2004; Widory 2007), whereas from the combustion of dung cakes and burning C3 plants were 13.4 to 15.5‰ (Agnihotri et al. 2011) and ~15‰ (Pavuluri et al. 2010), respectively. Martinelli et al. (2002) also reported the average levels of $\delta^{15}N$ as 13.2 ± 4.2‰ in PM emitted from both C3 and C4 plants. Sen et al. (2018) reported the higher $\delta^{15}N_{TN}$ level at Lucknow (10.5 ± 1.3‰), followed by Delhi (9.9 ± 2.6‰) and then least in Kolkata (9.8 ± 0.6‰) (Table 2). However, the $\delta^{15}N_{TN}$ level in PM_{10} at Bhubaneswar, Giridih, and Ajmer were recorded as 11.0 ± 2.1‰, 10.5 ± 0.8‰ and 10.5 ± 2.0‰, respectively, which are located outside of IGP, India (Sen et al. 2018).

The $\delta^{15}N_{TN}$ levels in PM_{10} at Delhi was reported as 9.6 ± 2.8‰ by Sharma et al. (2015), which was quite similar as observed by Sen et al. (2018), whereas the $\delta^{15}N_{TN}$ levels in Kolkata (7.4 ± 2.7‰) were 2.4‰ lower than those reported by Sen et al. (2018). Over the India Himalayan Region (IHR), the $\delta^{15}N_{TN}$ values at Mohal-Kullu (8.3 ± 1.2‰) were slightly higher than Darjeeling (8.7 ± 1.1‰). The average value of $\delta^{15}N_{TN}$ over IGP (10.1 ± 1.5‰) was ~1.5‰ higher than the IHR (8.6 ± 1.2‰) but lower than Bhubaneswar, Giridih, and Ajmer (Table 1). The $\delta^{15}N_{TN}$ levels in PM at Goa, a coastal site near to the Arabian Sea was reported as 10.8 ± 2.2‰, which was slightly higher (0.1 to 1.0‰) than the $\delta^{15}N_{TN}$ levels as reported in the IGP region (Sen et al. 2018; Agnihotri et al. 2015). It is to be noted that previous $\delta^{15}N_{TN}$ values recorded in Chennai by Pavuluri et al. (2010) and Mumbai by Aggarwal et al. (2013) were more than twice as observed by Sen et al. (2018).

Table 1. The comparison of $\delta^{13}C$ and $\delta^{15}N$ levels of $PM_{2.5}$ and PM_{10} collected at different sites of India and the world

Location	**PM**	**$\delta^{13}C$ (‰)**		**$\delta^{15}N$ (‰)**		**Reference**
		Range	**Mean ± SD**	**Range**	**Mean ± SD**	
Delhi, India (winter)	$PM_{2.5}$	-26.4, -23.4	-25.3 ± 0.5	3.3, 14.3	8.9 ± 2.1	Sharma et al., 2022
Delhi, India (summer)	$PM_{2.5}$	-26.7, -25.3	-26.1 ± 0.4	2.8, 11.5	6.4 ± 2.5	Sharma et al., 2022
Delhi, India (winter)	PM_{10}	-26.2, -23.8	-25.2 ± 0.6	3.4, 14.8	9.2 ± 2.5	Sharma et al., 2022
Delhi, India (summer)	PM_{10}	-26.8, -23.3	-25.5 ± 0.9	2.8, 15.5	8.8 ± 3.6	Sharma et al., 2022
Delhi, India	PM_{10}	-26.4, -24.8	-25.5 ± 0.5	3.3, 14.3	9.6 ± 2.8	Sharma et al., 2015
Varanasi, India	PM_{10}	-26.4, -23.3	-25.4 ± 0.8	2.8, 11.0	6.8 ± 2.4	Sharma et al., 2015
Kolkata, India	PM_{10}	-26.6, -24.9	-26.0 ± 0.4	2.8, 11.5	7.4 ± 2.7	Sharma et al., 2015
Chennai, India	PM_{10}	-29.5, -24.3	-25.4 ± 1.3	-	-	Pavuluri et al., 2010
Lucknow, India	PM_{10}	-26.0, -23.9	-24.8 ± 1.0	8.2, 12.1	10.5 ± 1.3	Sen et al., 2018
Kullu, India	PM_{10}	-26.5, -24.6	-25.9 ± 0.6	7.0, 9.6	8.3 ± 1.2	Sen et al., 2018
Darjeeling, India	PM_{10}	-26.5, -24.4	-25.6 ± 0.9	7.5, 10.1	8.7 ± 1.1	Sen et al., 2018
Xiamen, China	PM_{10}	-27.2, -25.4	-	5.5, 12.5	-	Chen et al., 2017
Wroclaw, Poland	PM_{10}	-26.9, -25.1	-26.1 ± 0.1	5.0, 13.7	9.9 ± 2.0	Gorka et al., 2012
Baoji, China	PM_{10}	-24.4, -22.5	-	8.8, 23.1	-	Wang et al., 2010
Mexico City, Mexico	PM_{10}	-29.5, -24.3	-25.4 ± 1.3	-	-	Lopez-Veneroni., 2009
Paris, France	PM_{10}	-26.7, -24.5	-26.2 ± 0.5	5.3, 16.1	10.7 ± 3.1	Widory et al., 2004; 2007
Barcelona, Spain	PM_{10}	-	-25.8 ± 0.4	-	-	Mari et al., 2016
Cienfuegos, Cuba	PM_{10}	-27.5, -19.3	-24.8 ± 1.2	1.5, 19.1	9.2 ± 4.4	Gomez et al., 2018

Relationship of Stable C and N Isotopes and Related Species

The concentrations and statistical analysis, as well as receptor models, study of carbonaceous species (OC, EC, and TC) of PM_{10} at Delhi, Varanasi, and Kolkata of IGP, India, revealed that the vehicular traffic emissions, fossil fuel combustion, and biomass burning (including wood-burning) are the significant sources of PM_{10} (Sharma et al. 2015). The positive linear relationship between OC and EC (Table 2) of PM_{10} over Delhi (r^2 = 0.78), Varanasi (r^2 = 0.88), and Kolkata (r^2 = 0.81) of IGP reveal that the abundance of carbonaceous aerosols over the IGP region of India (Sharma et al., 2015; 2021). Similarly, Ram and Sarin (2011) reported a relationship between OC and EC (r^2 = 0.66) and OC and K^+ (r^2 = 0.59) of PM_{10} at Kanpur, IGP of India, and demonstrated the pollutants emitted from the biomass burning are the significant sources of PM. Whereas, a weak positive correlation between $\delta^{13}C$ and TC was 0.27, 0.17, and 0.46 at Delhi, Varanasi, and Kolkata, indicating the role of inorganic C in ambient samples of PM_{10} (Table 2).

Table 2. Correlation matrix for total C & N, and their isotopes $\delta^{13}C$, $\delta^{15}N$ over IGP India

Parameters	Delhi	Varanasi	Kolkata
$\delta^{13}C$ *vs.* $\delta^{15}N$	0.15	0.31	0.42
$\delta^{13}C$ *vs.* TC	0.27	0.17	0.46
$\delta^{15}N$ *vs.* TN	0.34	0.097	0.63[a]
δ^{13} OC *vs.* OC	0.04	0.20	0.48[a]
TC *vs.* OC	0.86[a]	0.47[a]	0.75[a]
$\delta^{15}N$ *vs.* NH_4^+	0.27	–0.13	–0.06
$\delta^{15}N$ *vs.* NO_3^-	0.10	–0.04	0.25
TN *vs.* NH_4^+	0.56[a]	0.85[a]	0.36
TN *vs.* NO_3^-	0.64[a]	0.91[a]	0.74[a]

[a] Significant at $P < 0.05$.

Nitrogen species (NO_3^- and NH_4^+) in particulate matter arise mainly from secondary aerosols in the form of nitrate and ammonium (Seinfield and Pandis, 1998; Sharma et al. 2014b, c). The roadside vehicles, industrial and agricultural activities over this region are contributed to nitrogen species in ambient aerosols through oxidation of gaseous pollutants like NO and NO_2. In the ambient air, nitrogenous compounds like NH_4NO_3 and $(NH_4)_2SO_4$) are formed mainly from gas-to-particle conversion and neutralization process of NH_3 gas with atmospheric acid gases (Seinfield and Pandis 1998). Therefore, introducing N to atmospheric aerosols may involve a complex chemical process, resulting in a vast range of $\delta^{15}N$ (-15 to +30‰). Sharma et al. (2015) reported the positive correlation between $\delta^{15}N$ and TN of PM_{10} at Delhi, Varanasi, and Kolkata of IGP, India, which indicates the abundance of NH_4NO_3 and $(NH_4)_2SO_4$ over the region. The positive correlation between TN & NO_3^- and TN & SO_4^{2-} were recorded over sites, indicating the abundance of secondary inorganic aerosols in PM_{10} (Table 2). Pavuluri et al. (2010) has reported the average value $\delta^{15}N$ as 23.9 ± 3.3‰ (range: 15.7 to 32.2‰) in PM samples at Chennai, India, and reported the source of PM as local animal excreta and biofuel/biomass burning. Several researchers also reported similar results (Widory 2004; 2007; Bikkina et al. 2016; Bosch et al. 2014; Kelly et al. 2005).

Conclusion

In this chapter, we reported the stable carbon and nitrogen composition ($\delta^{13}C_{TC}$ and $\delta^{15}N_{TN}$) of aerosols ($PM_{2.5}$ and PM_{10}) used for source identification at several polluted urban regions of the country. The stable carbon ($\delta^{13}C_{TC}$) and nitrogen ($\delta^{13}N_{TN}$) isotopic analysis of aerosols over the most polluted IGP regions of India revealed that FFC, BB (dung cake, C3 & C4 plant matter and wood, etc.), and secondary inorganic aerosols are the sources of aerosol over the region. The levels of stable carbon ($\delta^{13}C_{TC}$) and nitrogen ($\delta^{13}N_{TN}$) detected in aerosols over IHR of India reveal that biofuel burning (crop residue and wood-burning), FFC (coal, diesel, and gasoline, etc.), and waste incineration are the major sources of aerosols. It is to be noted that the chemical tracers are non-conservative, and their ratios are likely to change significantly during long-distance transport due to changes in concentrations.

Acknowledgments

The authors thankfully acknowledge the Director, CSIR-NPL and Head, ES&BMD of CSIR-NPL, New Delhi for their encouragement and support for the study. The authors also acknowledge the CSIR, New Delhi for financial support for this study.

References

Aggarwal SG, Kawamura K, Umarji GS, Tachibana E, Patil RS, Gupta, PK (2013) Organic and inorganic markers and stable C-, N-isotopic compositions of tropical coastal aerosols from megacity Mumbai: sources of organic aerosols and atmospheric processing. *Atmos. Chem. Phys*. 13(9):4667-4680.

Agnihotri R, Mandal TK, Karapurkar SG, Naja M, Gadi R, Ahammmed YN, Kumar A, Saud T & Saxena M (2011) Stable carbon and nitrogen isotopic composition of bulk aerosols over India and northern Indian Ocean. *Atmos. Environ*. 45: 2828-2835.

Bikkina S, Kawamura K, Sarin M (2016) Stable carbon and nitrogen isotopic composition of fine mode aerosols ($PM_{2.5}$) over the Bay of Bengal: impact of continental sources. *Tellus* B 68.

Bosch C, Andersson A, Kirrillova, EN, Budhavant K, Tiwari S, Praveen PS, Russell LM, Beres ND, Ramanathan V, Gustafsson O (2014) Source-diagnostic dual isotope composition and optical properties of water soluble organic carbon and elemental carbon in the South Asian outflow intercepted over Indian ocean. *J. Geophys. Res. Atmos*. D-022127: 11743-11759.

Cachier H, Buat-Menard P, Fontugne M, Chesselet R (1986) Long-range transport of continentally-derived particulate carbon in the marine atmosphere: evidence from stable isotope studies. *Tellus B* 38(3-4):161-177.

Chen Y, Du W, Chen J, Hong Y, Zhao J, Xu L, Xiao H, (2017) Chemical composition, structural properties, and source apportionment of organic macro molecules in atmospheric PM_{10} in a coastal city of Southeast China. *Environ. Sci. Pollut. Res*. 24:5877-5887.

Chow JC, Watson JG, Chen LWA, Arnott WP, Moosmuller H (2004) Equivalence of elemental carbon by thermal/optical reflectance and transmittance with different temperature protocols. *Environ. Sci. Technol*. 38: 4414-4422.

Court JD, Goldsack JR, Ferrari LM, Polach HA (1981) The use of carbon isotopes in identifying urban air particulate sources. *Clean Air* 6-11.

Dockery DW, Pope CA, 1994. Acute respiratory effects of particulate air pollution. *Annual Review of Public Health* 15, 107-132.

Fisseha R, Saurer M, Jäggi M, Siegwolf RT, Dommen J, Szidat S, Samburova V, Baltensperger U (2009) Determination of primary and secondary sources of organic acids and carbonaceous aerosols using stable carbon isotopes. *Atmos. Environ*. 43(2):431-437.

Gomez YM, Satamaria, JM, Elustondo D, Hernandez CMA, Widory D (2018) Carbon and nitrogen isotopes unravels sources of aerosol contamination at Caribbean rural and urban coastal sites. *Sci. Tot. Environ*. 642:723-732.

Gorka M, Zwolinska E, Malkiewicz M, Lewicka-Szczebak D, Jedrysek MO, (2012) Carbon and nitrogen isotope analyses coupled with palynological data of PM_{10} in Wroclaw city (SW Poland)-assessment of anthropogenic impact. *Isot. Environ. Health Stud*. 48:327-344.

Gustafsson Ö, Kruså M, Zencak Z, Sheesley RJ, Granat L, Engström E, Praveen PS, Rao PSP, Leck C, Rodhe H (2009) Brown clouds over South Asia: biomass or fossil fuel combustion? *Science* 323(5913):495-498.

Hegde P, Kawamura K, Joshi H, Naja M (2016) Organic and inorganic components of aerosols over the central Himalayas: winter and summer variations in stable carbon and nitrogen isotopic composition. *Environ. Sci. Pollut. Res.* 23(7):6102-6118.

Jacobson MZ (2001) Control of fossil fuel particulate black carbon and organic matter, possibly the most effective method of slowing global warming. *J. Geophys. Res.* 107(19): 4410.

Jain S, Sharma SK, Mandal TK, Shenoy DM, Bardhan P, Srivastava MK, Chatterjee A, Saxena M (2016) Seasonal variations of stable carbon and nitrogen isotopic composition of PM_{10} at urban sites of Indo-Gangetic Plains (IGP) of India. *IASTA Bulletin* 22:597-601.

Kawamura K, Kobayashi M, Tsubonuma N, Mochida M, Watanabe T, Lee M (2004) Organic and inorganic compositions of marine aerosols from East Asia: seasonal variations of water-soluble dicarboxylic acids, major ions, total carbon and nitrogen, and stable C and N isotopic composition. In: Hill RJ, Kaplan IR (Eds.), Geochemical Investigations in earth and space science: A Tribute to Isaac R. Kaplan. *Geochem. Soc.* Publ no. 9: 243-265.

Kelly SD, Stein C, Jickells TD (2005) Carbon and nitrogen isotopic analysis of atmospheric organic matter. *Atmos. Environ.* 39: 6007-6011.

Kim BM, Park JS, Kim SW, Kim H, Jeon H, Cho C, Kim JH, Hong S, Rupakheti M, Panday AK, Park RJ, Hong J, Yoon SC, (2015) Source apportionment of PM_{10} mass and particulate carbon in the Kathmandu Valley, Nepal. *Atmos. Environ.* 123:190-199.

Liousse C, Penner JE, Chuang C, Walton JJ, Eddleman H, Cachier H (1996) A global three-dimensional model study of carbonaceous aerosol. *J. Geophy. Res.* 101: 19411-19432.

Lopez-Veneroni D (2009) The stable carbon isotope composition of $PM_{2.5}$ and PM_{10} in Mexico City Metropolitan Area air. *Atmos. Environ.* 43: 4491-4502.

Mari M, Sanchez-Soberon F, Audi-Miro C, van Drooge BL, Soler A, Grimalt JO, Schuhmacher M, (2016) Source apportionment of inorganic and organic PM in theambient air around a cement plant: assessment of complementary tools. *Aerosol Air Qual.* Res. 16:3230-3242.

Martinelli LA, Camargo PB, Lara LBLS, Victoria RL, Artaxo P (2002) Stable carbon and nitrogen isotope composition of bulk aerosol particles in a C4 plant landscape of southeast Brazil. *Atmos. Environ.* 36: 2427-2432.

Martinelli LA, Camargo PB, Lara LBLS, Victoria RL, Artaxo P (2002) Stable carbon and nitrogen isotopic composition of bulk aerosol particles in a C4 plant landscape of southeast Brazil. *Atmos. Environ.* 36(14):2427-2432.

Miyazaki Y, Fu PQ, Kawamura K, Mizoguchi Y, Yamanoi K (2012) Seasonal variations of stable carbon isotopic composition and biogenic tracer compounds of water-soluble organic aerosols in a deciduous forest. *Atmos. Chem. Phys.* 12:1367-1376.

Miyazaki Y, Kawamura K, Jung J, Furutani H, Uematsu M (2011) Latitudinal distributions of organic nitrogen and organic carbon in marine aerosols over the western North Pacific. *Atmos. Chem Phys.* 11(7):3037-3049.

Mkoma SL, Kawamura K, Tachibana E, Fu P (2014) Stable carbon and nitrogen isotopic compositions of tropical atmospheric aerosols: sources and contribution from burning of C3 and C4 plants to organic aerosols. *Tellus B 66.*

Moore H (1977) The isotopic composition of ammonia, nitrogen dioxide and nitrate in the atmosphere. *Atmos. Environ.* 11: 1239-1243.

Naru D, Harmelin-Vivien M, Gomoiu MT, Onciu TM, (2007) Influence of the Danube River inputs on C and N stable isotope ratios of the Romanian coastal waters and sediment (black sea). *Mar. Pollut. Bull.* 54, 1385-1394.

Paatero P (1997) Least squares formulation of robust non-negative factor analysis. *Chemometrics and Intelligent Laboratory Systems* 37(1):23-35.

Paatero P, Tapper U (1994) Positive matrix factorization: A non-negative factor model with optimal utilization of error estimates of data values. *Environmetrics* 5(2):111-126.

Pavuluri CM, Kawamura K, Fu PQ (2015) Atmospheric chemistry of nitrogenous aerosols in northeastern Asia: biological sources and secondary formation. *Atmos. Chem. Phys.* 15(17):9883-9896.

Pavuluri CM, Kawamura K, Swaminathan T, Tachibana E (2011) Stable carbon isotopic compositions of total carbon, dicarboxylic acids and glyoxylic acid in the tropical Indian aerosols: Implications for sources and photochemical processing of organic aerosols. *J. Geophys. Res. Atmos.* 116 (D18).

Pavuluri CM, Kawamura K, Tachibana E, Swaminathan T (2010) Elevated nitrogen isotope ratios of tropical Indian aerosols from Chennai: implication for the origins of aerosol nitrogen in South and Southeast Asia. *Atmos. Environ.* 44: 3597-3604.

Peng TR, Liang WJ, Liu TS, Lin YW, Zhan WJ, (2015) Assessing the authenticityof commercial deep-sea drinking water by chemical and isotopic approaches. *Isot. Environ. Health Stud.* 51:322-331.

Pope III C A, Ezzati M, Dockery D W (2009) Fine-particulate air pollution and life expectancy in the United States. *New England Journal of Medicine*, 360(4), 376-386.

Ram K, Sarin M M (2011) Day-night variability of EC, OC, WSOC and inorganic ions in urban environment of Indo-Gangetic Plain: Implications to secondary aerosol formation. *Atmos. Environ.* 45: 460-468.

Rengarajan R, Sarin MM, Sudheer AK (2007) Carbonaceous and inorganic species in atmospheric aerosols during wintertime over urban and high-altitude sites in North India. *J. Geophys. Res.* 112: D21307.

Russell KM, Galloway JN, Macko SA, Moody JL, Scudlark JR (1998) Sources of nitrogen in wet deposition to the Chesapeake Bay region. *Atmos. Environ.* 32: 2453-2465.

Seinfeld JH, Pandis SN (1998) Atmospheric chemistry and physics: from air pollution to climate change. Wiley, New York.

Sen A, Karapurkar SG, Saxena M, Shenoy DM, Chaterjee A, Choudhuri AK, Das T, Khan AH, Kuniyal JC, Pal S, Singh DP, Sharma SK, Kotnala RK, Mandal TK (2018) Stable isotopic composition (C & N) of PM_{10} over Indo-Gangetic Plains (IGP), adjoining regions and Indo-Himalayan Range during a winter, 2014 campaign. *Environ. Sci. Poll. Res.* 25(26): 26279-26296.

Sharma SK, Agarwal P, Mandal TK, Karapurkar SG, Shenoy DM, Peshin SK, Gupta A, Saxena M, Jain S, Sharma A, Saraswati (2017). Study on ambient air quality of megacity Delhi, India during Odd-Even strategy. *Mapan*, 32(2), 155-165.

Sharma SK, Karapurkar SG, Shenoy DM, Mandal TK, (2022) Stable carbon and nitrogen isotopic characteristics of $PM_{2.5}$ and PM_{10} in Delhi, India. *J. Atmos. Chem.* (accepted).

Sharma SK, Kumar M, Rohtash, Gupta NC, Saraswati, Saxena M, Mandal TK (2014c) Characteristics of ambient ammonia over Delhi, India. *Meteol. Atmos. Physics*. 124: 67-82.

Sharma SK, Mandal TK, Saxena M, Rashmi, Rohtash, Sharma A, Gautam R (2014b) Source apportionment of PM_{10} by using positive matrix factorization at an urban site of Delhi, India. *Urban Climate* 10:656-670.

Sharma SK, Mandal TK, Saxena M, Rashmi, Rohtash, Sharma A, Saud A (2014a) Variation of OC, EC, WSIC and trace metals of PM_{10} in Delhi. *J. Atmos. Solar Terrest. Phys*. 113: 10-22.

Sharma SK, Mandal TK, Shenoy DM, Bardhan P, Srivastava MK, Chatterjee A, Saxena M, Saraswati, Singh BP, Ghosh SK (2015) Variation of stable carbon and nitrogen isotopes composition of PM_{10} over Indo Gangetic Plain of India. *Bull. Environ. Contamin. Toxicol.,* 95 (5), 661-669.

Smith BN, Epstein S (1971) Two categories of 13C/12C ratios for higher plants. *Plant Physiol*. 47(3):380-384.

Turekian VC, Macko S, Ballentine, Donna S, Robert J, Garstang M (1998) Causes of bulk carbon and nitrogen isotopic fractionations in the products of vegetation burns: laboratory studies. *Chem. Geol*. 152: 181-192.

Turekian VC, Macko S, Ballentine D, Swap RJ, Garstang M (1998) Causes of bulk carbon and nitrogen isotopic fractionations in the products of vegetation burns: laboratory studies. *Chem. Geol*. 152(1):181-192.

Ulbrich IM, Canagaratna MR, Zhang Q, Worsnop DR, Jimenez JL (2009) Interpretation of organic components from Positive Matrix Factorization of aerosol mass spectrometric data. *Atmos. Chem. Phys*. 9(9):2891-2918.

Venkataraman C, Habib G, Eiguren-Fernandez A, Miguel AH, Friedlander SK (2005) Residential biofuels in south Asia: carbonaceous aerosol emissions and climate impacts. *Science* 307:1454-1456.

Waked A, Favez O, Alleman L Y, Piot C, Petit JE, Delaunay T, Verlinden E, Golly B, Besombes JL, Jaffrezo J L, Leoz-Garziandia E (2014) Source apportionment of PM_{10} in a north-western Europe regional urban background site (Lens, France) using positive matrix factorization and including pri-mary biogenic emissions. *Atmos. Chem. Phys.* 14: 3325-3346.

Wang G, Xie M, Hu S, Gao S, Tachibana E, Kawamura K, (2010) Dicarboxylicacids, metals and isotopic compositions of C and N in atmospheric aerosols from in-land China: implications for dust and coal burning emission and secondary aerosol formation. *Atmos. Chem. Phys*. 10:6087-6096.

Widory D (2006) Combustibles, fuels and their combustion products: a view through carbon isotopes. *Combust Theor. Model*. 10(5):831-841.

Widory D (2007) Nitrogen isotopes: tracers of origin and processes affecting PM_{10} in the atmosphere of Paris. *Atmos. Environ*. 41: 2382-2390.

Widory D, Roy S, Moullec Y, Le Goupil G, Cocherie A, Guerrot C (2004) The origin of atmospheric particles in Paris: a view through carbon and lead isotopes. *Atmos. Environ*. 38: 953-961.

Yeatman SG, Spokes LJ, Dennis PF, Jickells TD (2001) Comparisons of aerosol nitrogen isotopic composition at two polluted coastal sites. *Atmos. Environ*. 35: 1307-1320.

About the Editor

Dr. Binoy K. Saikia, PhD
Coal & Energy Division
CSIR-North East Institute of Science & Technology,
Jorhat, Assam, India
Email: bksaikia@neist.res.in
ORCID ID: 0000-0002-3382-6218

Dr. Binoy K. Saikia is a Principal Scientist at Coal & Energy Research Division, CSIR-North East Institute of Science and Technology, Jorhat and recipient of the "Shanti Swarup Bhatnagar Prize-2021", the highest Indian Science & Technology award, for his outstanding contributions in the field of earth and atmospheric sciences. He has more than 130 research papers, 50 conference papers, 10 books, and 5 patents in his credit. He has handled multiple R&D projects in the field of energy & environment. He has also served as board member of reputed journals and scientific committee.

Index

A

aerosol particles, vii, 2, 22, 44, 47, 48, 50, 52, 57, 62, 112, 142
aerosol(s), v, vi, vii, viii, ix, x, 1, 2, 3, 4, 5, 6, 9, 12, 13, 14, 16, 17, 18, 19, 20, 21, 22, 24, 43, 44, 45, 46, 47, 48, 49, 50, 51, 52, 54, 55, 56, 57, 58, 59, 62, 63, 67, 78, 79, 80, 81, 82, 83, 84, 85, 86, 87, 88, 89, 90, 91, 92, 94, 96, 97, 98, 99, 100, 101, 102, 104, 105, 106, 108, 109, 110, 111, 112, 113, 114, 115, 117, 118, 119, 122, 124, 126, 127, 128, 129, 130, 131, 133, 134, 136, 139, 140, 141, 142, 143, 144
airborne microplastic, 27, 34, 40
air-pollution, 62
atmosphere, v, vii, viii, ix, 1, 2, 3, 4, 5, 9, 10, 12, 13, 14, 16, 17, 18, 21, 22, 23, 24, 25, 26, 27, 28, 36, 37, 43, 44, 47, 48, 49, 50, 51, 52, 53, 55, 56, 58, 62, 63, 65, 66, 69, 73, 75, 80, 81, 82, 84, 85, 86, 87, 88, 89, 90, 92, 94, 95, 98, 99, 107, 108, 110,112, 113, 114, 115, 123, 132, 134, 141, 143, 144
atmospheric aerosols, iii, v, vii, viii, x, 13, 36, 43, 44, 45, 46, 48, 52, 54, 56, 57, 59, 117, 129, 133, 140, 142, 143, 144

B

bacteria, ix, 62, 63, 65, 66, 67, 70, 73, 80, 81, 82, 83, 85, 90, 91, 92, 93, 94, 95, 98, 100, 102, 103, 106, 108, 109, 112, 113, 114, 115
bioaerosol(s), v, vii, viii, ix, 61, 62, 63, 64, 65, 66, 67, 68, 69, 70, 71, 72, 73, 74, 75, 76, 77, 78, 79, 80, 81, 82, 83, 84, 85, 86, 90, 91, 92, 93, 94, 95, 96, 98, 99, 100, 101, 102, 103, 104, 105, 106, 109, 110, 111, 112, 113
biodiversity, ix, 56, 61, 62, 64
biomass, x, 9, 46, 47, 54, 86, 89, 90, 117, 119, 123, 125, 127, 129, 133, 134, 137, 139, 140, 141
bioterrorism, ix, 61, 62, 64, 79
bisphenol A, 29, 32, 46

C

cancer, 26, 32, 80, 101, 102, 104, 107, 109, 110, 111, 112, 113, 114
concentration-weight trajectory (CWT), 46, 55
Coronavirus, ix, 85, 86, 97, 115
COVID-19, ix, 85, 86, 97, 105, 106, 108, 110, 113, 115, 116
crystallinity, 29, 31, 33

E

ecological aspects, ix, 62, 64
economic, ix, 5, 62, 64
endotoxins, ix, 62, 63, 68, 91, 98, 102, 103
enrichment factor (EF), x, 15, 117, 118, 121, 122, 127
environmental pollution, 6, 12, 36, 38, 40

F

Fourier Transform Infra-Red (FT-IR), 35
fungi, ix, 62, 63, 65, 66, 69, 80, 81, 85, 90, 91, 92, 93, 94, 95, 96, 100, 103, 106, 109, 112, 113, 115

H

H1N1 virus, ix, 85, 86
humidity, 9, 49, 65, 66, 67, 72, 82, 94, 95

I

indoor environment, viii, 23, 24, 36, 81, 83, 92, 93, 105, 106, 112
infectious diseases, ix, 61, 64, 101, 102, 103, 105, 106, 116

L

lanthanum (La), 3, 8, 15, 16, 18, 31, 39
lutetium (Lu), 3, 8, 109, 111, 113, 116

M

Mass Closure (IMPROVE Model), 123
mass spectroscopy analysis, 35
microplastic(s) (MPs), v, vii, viii, 23, 24, 25, 26, 27, 28, 29, 33, 34, 35, 36, 37, 38, 39, 40, 41
mycotoxins, 68, 69, 91, 104

N

nanoplastics (NPs), 26, 40

O

organic aerosol (OA), 45, 46, 47, 48, 49, 50, 51, 52, 54, 55, 56, 57, 58, 59, 119, 132, 141, 142, 143
organic compounds, v, vii, viii, 4, 14, 18, 21, 43, 44, 45, 47, 48, 50, 52, 53, 54, 56, 58
oxygenated-PAHs (OPAHs), 46, 54

P

pandemic, ix, 85, 86, 90, 105
particle shape, 31, 33
particle size, 30, 31, 35, 58, 71, 79, 101
particulate matter (PM), vii, viii, x, 2, 4, 9, 11, 12, 13, 16, 20, 39, 40, 43, 44, 45, 58, 59, 62, 107, 110, 117, 119, 120, 125, 126, 128, 130, 131, 132, 134, 135, 137, 138, 139, 142
persistent organic pollutants (POPs), 38, 53, 54, 58
plastic-related toxicity, 32
PM_{10}, x, 12, 16, 117, 118, 119, 130, 131, 132, 133, 134, 135, 136, 137, 138, 139, 140, 141, 142, 143, 144
$PM_{2.5}$, x, 8, 9, 12, 14, 15, 16, 18, 19, 20, 22, 46, 54, 58, 59, 117, 118, 119, 121, 122, 123, 124, 125, 126, 127, 129, 130, 131, 132, 133, 134, 135, 136, 138, 140, 141, 142, 143
pollen, ix, 63, 65, 66, 68, 70, 85, 90, 91, 92, 94, 96, 102, 103, 109, 115
polychlorobiphenys (PCBs), 2, 54
polycyclic aromatic hydrocarbons (PAHs), 2, 46, 54, 57, 58
positive matrix factorization (PMF), x, 55, 117, 118, 119, 120, 121, 126, 127, 130, 131, 132, 144
potential source contribution function (PSCF), 55
principal component analysis (PCA), x, 15, 57, 117, 118, 119, 120, 121, 124, 125, 127, 130, 132
Promethium (Pm), 3, 8
protection and control methods, 62, 64

R

Raman Spectroscopy (RS), 35, 141
rare earth elements (REEs), v, vii, 1, 2, 3, 4, 5, 6, 7, 8, 9, 10, 11, 12, 13, 14, 15, 16, 17, 18, 19, 20, 21, 22
respiratory diseases, 11, 101, 102, 105

S

SARS-CoV-2, 67, 81, 97, 105, 111
Scandium (Sc), 3, 8
source apportionment, vi, vii, x, 20, 59, 117, 118, 119, 120, 128, 129, 130, 131, 134, 141
surface area, 12, 27, 29, 31, 40, 52
surface chemistry, 33, 38

Suspended Atmospheric Microplastics (SAMP), viii, 23, 24, 39

T

temperature, 39, 51, 65, 66, 68, 72, 81, 94, 95, 98, 110, 141
trace elements (TEs), v, vii, 1, 2, 3, 4, 5, 6, 7, 9, 10, 11, 12, 13, 14, 15, 16, 17, 18, 20
trace metals, 2, 3, 5, 13, 131, 144

U

Ultraviolet Germicidal Irradiation method (UVGI), 76
ultraviolet radiation, 82, 94
UNMIX, x, 117, 118, 119, 120, 121, 125, 127, 130, 132

V

viruses, ix, 62, 63, 65, 66, 67, 70, 73, 80, 85, 90, 91, 94, 96, 97, 100, 103, 104, 106, 109, 110

W

World Health Organization (WHO), ix, 12, 85, 97, 105, 115

Y

Yttrium (Y), 3, 8

β

β-glucans, 68

δ

$\delta^{13}C_{TC}$ a, x, 133, 134, 137, 140
$\delta^{13}T_{TN}$, x, 133